Sustainable Development Goals Series

The **Sustainable Development Goals Series** is Springer Nature's inaugural cross-imprint book series that addresses and supports the United Nations' seventeen Sustainable Development Goals. The series fosters comprehensive research focused on these global targets and endeavours to address some of society's greatest grand challenges. The SDGs are inherently multidisciplinary, and they bring people working across different fields together and working towards a common goal. In this spirit, the Sustainable Development Goals series is the first at Springer Nature to publish books under both the Springer and Palgrave Macmillan imprints, bringing the strengths of our imprints together.

The Sustainable Development Goals Series is organized into eighteen subseries: one subseries based around each of the seventeen respective Sustainable Development Goals, and an eighteenth subseries, "Connecting the Goals", which serves as a home for volumes addressing multiple goals or studying the SDGs as a whole. Each subseries is guided by an expert Subseries Advisor with years or decades of experience studying and addressing core components of their respective Goal.

The SDG Series has a remit as broad as the SDGs themselves, and contributions are welcome from scientists, academics, policymakers, and researchers working in fields related to any of the seventeen goals. If you are interested in contributing a monograph or curated volume to the series, please contact the Publishers: Zachary Romano [Springer; zachary.romano@springer.com] and Rachael Ballard [Palgrave Macmillan; rachael.ballard@palgrave.com].

Vedat Yorucu • Festus Victor Bekun
Ibrahim Yitmen

Modern Approach of Resilient and Sustainable Smart Cities

Springer

Vedat Yorucu
Department of Economics
Eastern Mediterranean University
Famagusta, Cyprus

Festus Victor Bekun
Department of Logistics Management
İstanbul Gelişim University
İstanbul, Türkiye

Ibrahim Yitmen
Department of Construction Engineering
and Lighting Science
Jönköping University
Jönköping, Sweden

ISSN 2523-3084 ISSN 2523-3092 (electronic)
Sustainable Development Goals Series
ISBN 978-3-032-05714-3 ISBN 978-3-032-05715-0 (eBook)
https://doi.org/10.1007/978-3-032-05715-0

This Springer imprint is published by the registered company Springer Nature Switzerland AG
The registered company address is: Gewerbestrasse 11, 6330 Cham, Switzerland

Preface

The rapid evolution of urban landscapes has necessitated a paradigm shift in how cities are planned, governed, and sustained. *Modern Approach of Resilient Sustainable Smart Cities* emerges as a critical exploration of this transformation, offering a comprehensive analysis of the technological, governance, and socio-economic dimensions shaping the future of urbanization.

The book begins by contextualizing the rise of smart cities within the broader historical trajectory of urban development—from the Industrial Revolution to today's digitally interconnected metropolises. Urbanization has always been a driver of innovation, yet contemporary challenges—climate change, infrastructural decay, economic disparities, and technological disruption—demand a reimagined approach. Smart cities, as this book argues, are not merely about digital integration but about resilience, sustainability, and equitable governance.

The first section delves into the conceptual foundations of smart cities, dissecting varying definitions and implementation models. While IBM's data-driven optimization and Amsterdam's sustainability-focused initiatives exemplify divergent approaches, the book underscores a critical reality: smart city frameworks are not universally replicable. Disparities in technological access, financial capacity, and governance structures create uneven playing fields, particularly for developing economies. Case studies from Singapore, Barcelona, and Nairobi reveal both the promises and pitfalls of smart urbanism—highlighting successes while cautioning against technocratic solutions that neglect social equity.

A central theme is the interplay between technology and governance. Smart infrastructure, IoT networks, and AI-driven analytics are meaningless without robust policy frameworks and participatory decision-making. The book critically examines governance models—centralized, decentralized, and hybrid—weighing their efficiencies against risks of exclusion, corporate overreach, and cybersecurity threats. It challenges the assumption that technological advancements alone can solve urban crises, advocating instead for ethical frameworks that prioritize privacy, inclusivity, and democratic engagement.

The discussion then shifts to the stark contrast between smart city development in advanced and emerging economies. While cities like London and Copenhagen leverage AI and 5G to optimize services, regions in Africa and South Asia grapple with infrastructural gaps and political instability. The book critiques the uncritical transplantation of Western models, urging context-sensitive strategies that address local needs, frugal innovation, and systemic inequities.

Later chapters pivot to actionable resilience strategies, illustrated through the case study of Catalkoy-Esentepe in Northern Cyprus. Here, the book transitions from theory to practice, examining how smart energy grids, renewable transitions, and digital governance can revitalize urban systems. The analysis extends to spatial planning, waste management, and tourism-driven economies, emphasizing that resilience requires not just technology but holistic policy alignment and community participation.

The final chapters present a forward-looking vision for resilient, sustainable smart cities. An integrated framework combining AI-driven risk assessment, IoT monitoring, and adaptive governance is introduced to transform urban resilience from reactive to predictive. This is expanded through Urban Digital Twins (UDTs), demonstrating how virtual modeling can optimize everything from infrastructure to emergency response while aligning with SDG 11. Finally, these concepts are grounded in global realities, contrasting European green initiatives, Moscow's sustainable redevelopment, and India's smart city program to highlight context-sensitive solutions. The study compares smart city business models across London, Amsterdam, and Berlin, revealing key differences in financing mechanisms, public-private partnerships, and community-centered urban innovation.

This book is the product of collaborative scholarship, blending academic rigor with pragmatic solutions. It is intended for urban planners, policymakers, technologists, and activists who seek to navigate the complexities of twenty-first-century urbanization. As cities worldwide confront unprecedented challenges—from climate crises to digital divides—this work serves as both a roadmap and a cautionary tale, reminding us that the future of urban living hinges not on technology alone, but on our collective commitment to justice, sustainability, and resilience.

Famagusta, Cyprus — Vedat Yorucu
İstanbul, Turkey — Festus Victor Bekun
Jönköping, Sweden — Ibrahim Yitmen

Acknowledgments

The successful completion of Modern Approach of Resilient Sustainable *Smart Cities* marks the culmination of a collaborative effort driven by shared expertise and dedication. As the authors of this work, we extend our deepest gratitude to all those who contributed to its realization. This book is the product of rigorous intellectual exchange, innovative thinking, and a collective commitment to advancing the discourse on resilient and sustainable urban development. Each contributor brought unique insights and specialized knowledge, enriching the multidisciplinary approach essential for addressing the complexities of smart cities. The synergy of our collaboration—built on mutual respect, critical dialogue, and a passion for transformative solutions—has been instrumental in shaping this work.

Several pioneers with academic expertise on resilience of sustainable smart cities have contributed to making this study possible, and it is our pleasure to acknowledge their contributions, while, of course, stating that all remaining errors of fact or interpretation are ours entirely. We owe special thanks to Professor Pınar Uluçay Righelato, Senior Lecturer of Eastern Mediterranean University, who has read our monographs and made extensive and constructive comments. We are privileged to express our profound gratitude to Assoc. Prof. Dr. Mary Oluwatoyin Agboola, coordinator of Graduate Studies and special Projects in Academic Affairs Howard Payne University, Brownwood, Texas USA for her constructive input and support during the peer review of this book. We would like to express our deepest gratitude to Prof. Lavinia Chiara Tagliabue from the University of Turin for her invaluable feedback, insightful suggestions, and enthusiastic support throughout the review process of this book. Her expertise and thoughtful critique have greatly enriched the quality of this work. We sincerely appreciate her time, dedication, and encouragement, which has been instrumental in shaping the final manuscript.

We are also sincerely thankful to the editors and publishers for their meticulous attention to detail, which ensured the manuscript's clarity, coherence, and polished presentation. Their dedication and professionalism have been indispensable in bringing this project to fruition. We also acknowledge the broader academic and professional community dedicated to sustainable urban innovation; your pioneering research continues to inspire our work.

On a personal note, we are profoundly grateful to our families for their unwavering support and patience throughout this journey. Their encouragement provided the motivation needed to overcome challenges and bring this vision to fruition. We dedicate this study to our wives, namely Saskia Yorucu, Emine Davut Yitmen, and Gloria Nnadwa Alhassan, respectively, for their love and support. We cannot complete this acknowledgment section without special thanks to the love of our children, Stefaan and Suzanne, Uras, and Festus Jr. M. Festus, correspondingly.

This book stands as a testament to the power of collective knowledge and forward-thinking collaboration. We hope it serves as a meaningful contribution to the ongoing global conversation on building resilient, sustainable, and inclusive cities for the future.

Competing Interests The authors have no competing interests to declare that are relevant to the content of this manuscript.

Contents

About the Authors

Vedat Yorucu was born in 1970 in Nicosia. He earned his Ph.D. in Economics from the University of Leicester in 1998. He has been working at Eastern Mediterranean University since 1998. In addition to his academic career, he served for 5 years as Chief Economic Advisor to the Minister of Economy and Finance and the President of Northern Cyprus. He also served as Visiting Scholar at the College of Europe-Brugge Campus (Belgium) during 2013–2017 summer and at Norman Paterson School of International Affairs-Carleton University, Ottawa, for 2018 summer. He has published many scholarly articles in leading international refereed journals in the field of economic development, construction economics, energy policy, energy economics, and applied econometrics. He is co-author of books published by Springer Publishing Inc.: *Southern Energy Corridor: The Role of Turkey in European Energy Security* (2018), *Modern Geopolitics of Eastern Mediterranean Hydrocarbons in an Age of Energy Transformation* (2020), *Islands in Maritime Disputes: Greece Turkey Geo-Political Relations* (2022), and *From Land Disputes to Sustainable Development: A Near East Perspective* (2024).

Festus Victor Bekun currently serves as an Associate Professor at Istanbul Gelisim University in Turkey, and he is also a senior adjunct research fellow at VIZJA University Warsaw, Poland. Dr. Bekun has over 311+ refereed scientific publications indexed in WOS and Scopus index top-tier outlet. Dr. Bekun's research interest is in the areas of agricultural economics, energy economics, sustainability, and carbon reduction policies. Dr.Bekun's works have appeared in top-tier journals including *Science of The Total Environment*, *Energy & Environment*, *Journal of Cleaner Production*, *Energy & Environment*, Sustainable Development, more recently Current Issues in *Tourism*, *Heliyon*, and the prestigious *Renewable and Sustainable Energy Review* (RSER) journal, among others. The author is currently serving as either Editor or Associate Editor to several WOS and SCOPUS index journals namely: Associate Editor—*Journal of Sustainable Finance and Investment*, Editor—*Open Agriculture*, Editor—*Journal of Risk and Financial Management*, Guest Editor—*Sustainability*, Associate Editor—*Journal of Public Affairs*, among others.

Ibrahim Yitmen holds a Ph.D. in Architecture from Istanbul Technical University, Turkey. Since February 2018, he has served as Professor of Construction Production Management at Jönköping University, Sweden, where he leads research at the forefront of digital innovation in the architecture, engineering, and construction (AEC) sector. His interdisciplinary work bridges socio-technical systems and digital transformation, with recent focus areas including Construction 5.0 for smart, sustainable, and resilient built environments, and Digital Twin-enabled cognitive buildings and infrastructure lifecycle management. A prolific scholar, Professor Yitmen, has authored 180+ peer-reviewed publications in high-impact journals and conference proceedings. His editorial leadership includes five seminal books on various topics including BIM-enabled Cognitive Computing, Cognitive Digital Twins, and Mars Architecture: Construction 6.0 for Sustainable and Health-Oriented Habitats. An active contributor to global AEC research, Professor Yitmen serves on scientific committees for IT and Computing in Construction at CIB, ASCE, and EC3 conferences and reviews for top-tier journals.

Abbreviations

5G	Fifth Generation Mobile Network
AEC	Architecture, Engineering and Construction
AI	Artificial Intelligence
AR	Augmented Reality
BIM	Building Information Modelling
BOT	Build Operate Transfer
BREEAM	Building Research Establishment Environmental Assessment Methodology
CADER	Catalkoyü Geliştirme ve Kültür Derneği (Association of Culture and Development of Catalkoy)
CIM	City Information Model
CRI	City Resilience Index
CSO	Civil Society Organizations
DT	Digital Twins
EGCN	European Green Capitals Network
EHCN	European Healthy Cities Network
EMF	Electromagnetic Field Exposure
EU	European Union
GCD	Geiger Counter Devices
GHG	Greenhouse Gases
GIS	Geographic Information Systems
ICLEI	International Council for Local Environmental Initiatives
ICT	Information and communication technology
IoT	Internet of Things
IRADe	Integrated Research and Action for Development
KIBTEK	Kıbrıs Türk Electrik Kurumu (Cyprus Turkish Electricity Authority)
LED	Lighting retrofits
LEED	Leadership in Energy and Environmental Design
LEZs	Low Emissions Zones
LLM	Large Language Models
MCR	Making Cities Resilient
MiC	Modular Integrated Construction
ML	Machine Learning
NB-IoT	Narrow-Band Internet-of-Things
NGO	Non-governmental organization
OECD	Organization for Economic Co-operation and Development

OS	Operating system
PPP	Public-Private Partnerships
RCN	Resilient Cities Network
RL	Reinforcement Learning
SC	Smart City
SCG	Smart Cities Governance
SCM	Smart Cities Mission
SCO	Civil Society Organization
SCWG	Smart Cities Working Group
SDGs	Sustainable Development Goals
SDT	Sustainable Development Theory
SDuMuRS	Sustainability, Durability, Multi-hazard resistance, Resilience, and Smart monitoring and management
TAM	Technology Acceptance Model
THM	Technology, Human, and Management
TRBOT	Toronto Region Board of Trade
UDT	Urban Digital Twins
UIA	International Union of Architects
ULEZ	Ultra Low Emission Zones
UN	United Nations
UNDRR	United Nations Office for Disaster Risk Reduction
VR	Virtual Reality
WEEE	Waste of Electric Electronic Equipment
WiFi	Wireless fidelity
WSN	Wireless Sensor Network
XR	Extended Reality
ZEZ	Zero-Emission Mobility Zones

1 Introduction

The rationale of this study emphasizes the transformation of traditional cities into resilient sustainable smart cities. Sustainable development is the necessity of modernization and innovation goals of the UN Sustainable Development Goals (UN SDG). From this perspective, this monograph has been written by three authors from different disciplines. It is a multidisciplinary work on the modern approach of resilient sustainable smart cities, offering a comprehensive analysis of the technological, governance, and socio-economic dimensions shaping the future of urbanization. The book begins by contextualizing the rise of smart cities within the broader historical trajectory of urban development from the Industrial Revolution to today's digitally interconnected metropolises. Urbanization has always been a driver of innovation, yet contemporary challenges, climate change, infrastructural decay, economic disparities, and technological disruption. Smart cities, as this book argues, are not merely about digital integration but about resilience, sustainability, and equitable governance.

Chapter 2 analyses the multifaceted concept of smart cities, tracing their historical emergence from post-industrial urbanization to contemporary technological innovations. Readers of this chapter will have access to several definitions of diversity of smart cities, acknowledging both the strengths and limitations of such variability in fostering local adaptability. Furthermore, this chapter will discuss the key components of smart cities, such as infrastructure, governance, sustainability, and technology integration, emphasizing the necessity of supportive regulatory frameworks. Moreover, smart city challenges such as data privacy, cybersecurity, financial constraints, and the digital divide will also be discussed in Chap. 2. Ultimately, while the chapter presents an optimistic view of smart cities' transformative potential, it concludes that a more critical interrogation of political, economic, and power structures is essential for equitable urban futures.

The third chapter of the tentative blueprint for smart cities will look at the role of governance in smart cities development and evolution. Chapter 3 also aims to outline how multi-stakeholders cooperate and work together to succeed in smart city strategies and practices. Moreover, Chap. 3 will examine the role of best governance in smart cities development or evolution. Besides, the role of shareholders in smart city governance is going to be examined in Chap. 3. Additionally, the civil society organizations in sustaining cities resilience as well as the key components of smart cities governance, such as policy frameworks, decision-making processes, regulatory framework and governance models in smart cities will be explored throughout in Chap. 3. There are several models available for best practice governance, which have been implemented by several successful smart cities, such as Barcelona, Copenhagen, and Amsterdam. The success sto-

V. Yorucu et al., *Modern Approach of Resilient and Sustainable Smart Cities*, Sustainable Development Goals Series, https://doi.org/10.1007/978-3-032-05715-0_1

ries in those cities will be discussed in Chap. 3. Furthermore, Chap. 3 underscored disparities in access to technology, governance, and finance, which complicate scalability and replicability.

Chapter 4 explores the dynamics of smart cities in developed and developing economies. Readers can reach information on the evolution of smart cities, the core components and factors to achieve goals and strategies to be a smart city. In this chapter, discussions to further discourse on urban transformation by offering strategic insights for policymakers, urban planners and technology stakeholders will be analyzed. Readers of Chap. 4 will have an idea about the smart city development processes within developed and developing economies, which generate specific implementation advantages and barriers. Moreover, the combination of advanced digital infrastructure, AI-driven governance, and high-speed connectivity has been shown to allow developed nations to manage their cities efficiently. Contrary to say, the emphasis of developing economies lies in mobile-first solutions combined with foreign investments and low-cost technological innovations to address their infrastructure shortcomings. Moreover, Chap. 4 will report how two economies struggle with identical problems, including data privacy issues, cybersecurity risks and socioeconomic inequalities. Likewise, Chap. 4 will investigate the uncritical transfer of smart city models from the Global North to the Global South, arguing for locally tailored approaches. It also seeks to analyze the implementation of emerging technologies like AI, IoT, and blockchain alongside their ethical implications. Chapter 4 also aims to expose information about robust infrastructure that combines smart technology with efficient transportation and sustainable waste management, which is vital for both developed and developing economies. Furthermore, developed nations should perfect sustainability practices, but developing countries need budget-friendly sustainable practices that work across regions to handle fast urban growth and environmental deterioration. Readers of Chap. 4 will reach the conclusion that the successful implementation of smart cities requires striking a balance between technological progress and equitable social support alongside clear governance practices and environmentally sustainable solutions. Consequently, Chap. 4 concludes by advocating for inclusive, adaptive, and ethically governed smart city frameworks that prioritize human development over purely technological advancement.

Literature on migration, population growth and threats for smart urban cities will be discussed in Chap. 5, which Catalkoy-Esentepe municipal region has been selected as a case study, including six villages within administrative boundaries. In Chap. 5, readers of this monograph will have detailed information about the population growth in the region for the past two centuries, covering the period of 1831–2024 within the perspective of urban mobility. Population growth can be the result of increasing migration from rural to urban cities, which reduces the quality of life, welfare level, endangering safety and security issues of the urban city. It is a threat for the resilience of the city. Rapid population growth has increased the need for housing, green and open spaces, and utility services. Major infrastructural and social problems, such as waste management, pollution, traffic congestion, and water and energy shortages in the region will be exposed within Chap. 5.

Additionally, readers of Chap. 5 will have comprehensive information about the needs of a smart city during the transformation process. Urban mobility causes an increasing number of motor vehicles in use, which causes severe traffic congestion, carbon emission, noise pollution, hazards of road accidents, and waste of financial resources on gasoline expenditures. To safeguard carbon free environment, it is proposed to expand bicycle roads and pedestrian paths. Chapter 5 also discusses how to improve public transportation in Catalkoy-Esentepe region. This chapter will explore the discussions about the concepts of rainwater management, sanitation, solid waste management, energy efficiency, rehabilitation of green and blue spaces, construction of advanced technology of internet infrastructure, good governance and the need for spatial planning for the development of Catalkoy-Esentepe as a case study of future smart city in North Cyprus.

Chapter 6 will shed light on the transformation process of traditional cities into a safe, inclusive, resilient and sustainable smart urban city. The historical, cultural and economic impact of city life has caused rapid urban mobility due to big potential of the city, in terms of economic benefits and employment. Throughout Chap. 6, readers of this study will have information about necessary action plans, policies and objectives needed to be implemented to succeed sustainability objectives of SDG 11. The region of Catalkoy-Esentepe has very rich historical heritage and archaeological sites to promote the region as a sustainable tourist region. Chapter 6 will highlight information on physical infrastructure capacity of tourism and construction. The region has nearly 50% of five-star hotels and enrolls in a significant amount of job opportunities in the service industry. Other than tourism, the construction sector boomed in the region during the last 10 years and the estate market became the locomotive of the regional income. Such economic developments have initiated urban mobility, which increased population growth, the demand for housing and labor force. Readers of this chapter will also have an idea about the impacts of negative externalities to city life, such as traffic congestion, increasing rents, insufficient housing, lack of utility services, pollution, climate change and infrastructure bottlenecks. This chapter shed light on the basic urgent needs of the city and necessary action plans to be considered to make the city safer and more resilient. The implementation of spatial planning, action plans, objectives, strategies, policy recommendations and the assessment needed to succeed the transformation process will be elaborated in the next two chapters.

Chapter 7 aims to examine the city level development plans, action plans, objectives, strategies and policy measures of sustainable urban cities to achieve SDG 11 objectives. The plans, programs and strategies are subject to be assessed by the key performance indicators to measure the performance of the urban city. Ultimately, readers of this chapter will find information specifically on urban development, housing, green and open spaces, waste management, blue spaces and water management, urban policy and governance. The urban development proceeded in the absence of environmental planning and a coherent approach of housing. Rapid urbanization development in Catalkoy-Esentepe has resulted in unplanned urban sprawl, substandard and unaffordable housing, traffic congestion, air and sound pollution, infrastructure bottlenecks, blackouts, unfriendly urban design, lack of green spaces and waste management problem. Over the last 10 years Catalkoy-Esentepe has made important progress in addressing these problems, however lack of finance, migration and population growth in the region, mismanagement by the local administrations and lack of knowledge resulted in living in a city without implementation of such urban development plans. The implementation of regional land zone plans has been examined with detailed cadastral maps in Chap. 7 and city action plans, objectives, strategies and policy measures which are required to be followed are explained over here. Moreover, Chap. 7 focuses on key performance indicators that need to be implemented to assess the city's successful performance.

In Chap. 8, readers of this monograph will find valuable information on the developmental significance and implications of smart and resilient cities, with a particular focus on the challenges inherent in their planning, design, and construction. Besides that, this chapter investigates the potential of artificial intelligence (AI) techniques to improve scenario planning methodologies within the context of smart and resilient urban development. Chapter 8 also addresses the technical and institutional challenges associated with AI integration and proposes practical solutions, while emphasizing the indispensable role of planning professionals in guiding the AI-enabled decision-making process. In addition, this chapter shed light on the objectives and foundational principles of territorial spatial resilience planning and proposes a comprehensive framework that incorporates technical, spatial governance, and operational dimensions. Furthermore, the significance of smart resilient cities lies in their ability to address critical needs in disaster prevention, urban governance, and resource allocation. They not only safeguard public safety during emergencies but also improve the quality of life for citizens by ensuring efficient public services, sustainable resource management, and equitable access to infrastructure. Furthermore, the integration of AI into scenario planning enhances the adaptability of urban systems, enabling cities to predict and mitigate the impacts of natural and human-made disasters more effectively. Moreover, a technology-driven framework for addressing the complexities of urbanization has been elaborated over here. Chapter 8 is a very detailed one, which introduces the emerging concept of Smart-Resilient 5.0 from an environmental perspective, outlining its theoretical foundations and identifying priority research areas that characterize the next evolution in smart and resilient urban systems. This unique study is novel in its field, which underlines the key components of smart and resilient urban development. Comprehensive assessment indicators, which were highlighted over here, are further guide cities in tracking progress, identifying weaknesses, and improving urban governance. Chapter 8 concludes with the discussion of how smart resilient cities represent the future of urban development, blending technological advancements with sustainability and inclusivity to create adaptive and sustainable environments. Through collaboration, innovation, and governance, these cities ensure resilience against future challenges while enhancing quality of life.

Chapter 9 focuses on sustainability, durability, multi-hazard resistance, resilience, and smart monitoring guides to the design of future urban infrastructures. This chapter shed light on urban resilience, defined through engineering, ecological, and socio-ecological perspectives, focuses on a city's capacity to withstand and adapt to various shocks. Readers will reach information about integrating Urban Digital Twins (UDT) into smart city frameworks, which has the potential to revolutionize urban development. The UDT emerge as pivotal tools in smart city development. These digital replicas of urban environments facilitate real-time interaction and decision-making, enhancing infrastructure management and sustainability efforts. The applications of UDT in urban planning are diverse, offering innovative solutions for energy management, infrastructure monitoring, and public services, among other areas. Chapter 9 also enlightens that many cities are exploring the concept of UDT to enhance urban administration, development, and decision-making processes. The UDT can enable faster response times through autonomous actions and enhance city management efficiency by providing real-time insights into the city's condition. Chapter 9 concludes with the discussion on the Construction 5.0 for resilient, sustainable and urban cities. It simply represents the next evolution in the architecture, engineering, and construction (AEC) sector, emphasizing human-centric technology, resilience, and sustainability. It integrates digital technologies like AI, IoT, and blockchain to enhance efficiency and reduce environmental impacts.

Chapter 10 reports a comprehensive analysis of sustainable urban resilience development initiatives, with a particular focus on their implementation across European cities. The municipal commitments to advancing environmental sustainability and public health, alongside efforts to strengthen local governance frameworks aligned with the Sustainable Development Goals (SDGs) are inspected within Chap. 10. The chapter also investigates the incorporation of green infrastructure and nature-based solutions as part of broader urban ecological strategies. It synthesizes lessons learned from past urban resilience initiatives, identifying critical insights for shaping future sustainable urban development. Several case studies, especially from Russia, China and India are examined and lessons from Russian (Moscow) experiences have been examined, which shows the role of green construction in promoting resilience, supported by urban planning regulations and risk-informed land-use strategies. Moreover, Chap. 10 examines emerging green construction standards and environmental certification systems in Russia, contextualized within global best practices. Chapter 10 ends with a comparative evaluation of smart city business models from London, Amsterdam, and Berlin, which provide additional insights into financing mechanisms, public-private partnerships, and community-centered urban innovation. The comparative study of smart city business models demonstrates how mature European cities are fine-tuning their resilience agendas by leveraging business models that align with local conditions, technologies, and policy goals.

Chapter 11 draws the major conclusions and highlights of the study. As a summary, this book reports on the fact that sustainable urban resilience is not a destination but an ongoing journey. It demands collaboration, innovation, and a steadfast commitment to shaping cities that are not only smart and efficient but also just livable and prepared for the uncertainties of tomorrow.

The Anatomy of Smart Cities: A Global Perspective

2

Abstract

The global trend towards smart city development is gaining momentum, particularly in countries with advanced scientific and environmental capabilities. As such, this chapter examines the multifaceted concept of smart cities, tracing their historical emergence from post-industrial urbanisation to contemporary technological innovations. It explores the definitional diversity of smart cities, acknowledging both the strengths and limitations of such variability in fostering local adaptability. Through comparative case studies from Singapore, Amsterdam, and other global regions, the chapter underscored disparities in access to technology, governance, and finance, which complicate scalability and replicability.

Furthermore, this chapter discussed key components of smart cities, such as infrastructure, governance, sustainability, and technology integration, emphasising the necessity of supportive regulatory frameworks. Moreover, smart city challenges such as data privacy, cybersecurity, financial constraints, and the digital divide were also discussed. Moreso, while highlighting emerging technologies like 5G, AI, and blockchain, the chapter also highlights the insufficient engagement with ethical and legal concerns surrounding surveillance. Ultimately, while the chapter presents an optimistic view of smart cities' transformative potential, it concludes that a more critical interrogation of political, economic, and power structures is essential for equitable urban futures.

2.1 Introduction

The world is rapidly transforming due to urbanisation. With the start of the Second Industrial Revolution in the mid-nineteenth century, cities became engines of economic growth by drawing the rural population, which needed industrial jobs. According to Reis et al. (2022), this migration further forced the growth of cities, forming highly concentrated production and consumption centres. Moreover, Alagirisamy and Ramesh (2022) emphasise that global population growth has had a direct impact on the rise and expansion of cities. Due to this shift, approximately 56% of the total global population of around 4.4 billion individuals currently reside in urban areas (Hussain 2024). It is projected that urbanisation will peak with approximately 70% of the world's population by 2050, a trend that studies have found will increase pressure on physical infrastructure, natural resources, and social systems (Liang et al. 2025).

Particularly, with a surge in urbanisation, cities have proved to encounter more complex and apparent issues. Among these are risks associated

V. Yorucu et al., *Modern Approach of Resilient and Sustainable Smart Cities*, Sustainable Development Goals Series, https://doi.org/10.1007/978-3-032-05715-0_2

with climate and energy insecurity (Heidari et al. 2022), environmental degradation and deterioration in living conditions (Reis et al. 2022), as well as housing shortages, healthcare disparities, traffic congestion, and urban sprawl (Sharifi et al. 2021). Traditional governance and planning mechanisms have not well managed this complexity. As part of that response, the smart city paradigm has not emerged as a panacea for city problems but rather as a data-driven, technology-enabled model that seeks to streamline city management, improve quality of life, and reduce resource inefficiencies. As Tanwar et al. (2022) explain, smart cities provide a transformational route to meet the multidimensional demands of modern urbanisation by integrating digital infrastructure, real-time analytics, and citizen involvement.

Moreover, the concept of smart cities is imprecise, with their significance varying across countries due to geography, ecosystem, resource availability, and significant challenges that need to be addressed. While there are many definitions of what constitutes a smart city, Hussain (2024) defines a smart city as a municipality that utilises information and communication technology to enhance operational efficiency, disseminate information to the public, and improve the quality of government services and citizen welfare. Building on this line, Chen et al. (2024) define a smart city as an urban region where technology and data collection contribute to a better quality of life, as well as the sustainability and efficiency of city operations. In its simplest manner, Huda et al. (2024) define a smart city as an intelligent, instrumented, and interconnected urban area. Collectively, these definitions conclude that a smart city is a deliberate pursuit of ICTs to change the way of life and work in a specific location.

Smart cities leverage advanced technologies and data-driven solutions to benefit various aspects of urban life, including infrastructure, transportation, governance, and sustainability. As pointed out by Javed et al. (2022), smart cities offer a vision of urban living that is environmentally friendly, adaptable to the changing circumstances of residents, and productive for society. Moreover, smart cities are envisioned as the cities of the future that address the problems faced by people living in traditional cities, such as traffic congestion, inadequate infrastructure, pollution, and inefficient resource management (Kozłowski and Suwar 2021). Smart cities utilise technology, data analytics, and networking to address urban issues, aiming to create a more comfortable, resilient, and resource-efficient urban environment through the interconnection of systems and services (Bhushan et al. 2020).

Although cities are often portrayed as a solution to local problems, smart cities promoted by governments around the world are local constructions that uniquely respond to socio-economic, cultural, and spatial contexts. Despite international organisations such as the OECD and the EU suggesting standardised models to encourage interoperability and scalability, these models may risk neglecting individual city requirements and capabilities. According to Moura and de Abreu e Silva (2021), successful smart city development requires a local level of adaptability rather than global standardisation. Accordingly, cities must customise their approaches to handle distinct challenges that can vary from infrastructure gaps to climate resiliency, aligning with overall sustainable and equitable objectives. Therefore, the smart city cannot be perceived as a rigid technological plan but rather as an adaptive, situational process of inclusive governance and planning in response to circumstances.

2.2 Importance of Smart Cities from a Global Perspective

The idea of the smart city is not merely a technological renovation of the urban environments; it is a game-changer in addressing the complex problems of rapid urbanisation, climate change, and socio-economic inequalities. In a global context, smart cities are being reappropriated as a means of improving the efficiency, sustainability, and quality of life in urban areas by integrating digital infrastructure, data analytics, and participatory governance. This international rhetoric contributes to a common language of creativity and strength, but the application is firmly grounded in the local circumstances. Cities worldwide are facing similar pressures, including, but not limited to, environmental degradation, resource shortages, and high population density (Butt 2021). However, solutions must take into account cultural, political, and geographic diversity (Butt 2021).

Comparative analyses across nations demonstrate that smart cities cannot be viewed as models but rather as differentiated adaptations to local urban environments. For example, Asian megacities tend to focus on alleviating congestion and air pollution, whereas the majority of European programs emphasise low-carbon infrastructure and community-based planning (Nastjuk et al. 2022). This diversity emphasises mutual learning and the strategic flow of best practices. International organisations and the scientific community should focus on the process of knowledge exchange and cross-cultural communication, along with joint experimentation, to tap into collective urban intelligence and prevent the development of redundant or inequitable solutions.

Notably, smart city research offers an effective vantage point for understanding the impacts of disparate governance frameworks and technological capabilities on urban development. As Heidari et al. (2022) point out, the effectiveness of smart city projects sometimes hinges not on the technology itself but on the compatibility between digital offerings and the social and political frameworks within a particular location. A global comparative analysis of successful projects and those that have failed offers important implications regarding institutional preparedness, citizen participation, regulatory environment, and scalability of infrastructure (Reis et al. 2022). These lessons are critical in guiding context-specific design and ensuring the long-term sustainability and inclusivity of innovative cities.

Furthermore, smart cities are part of a shared global city resolution-making. Cities are no longer operating in a vacuum but within more interconnected webs of practice. A combination of transdisciplinary research is necessary to address shared urban problems, such as climate change, demographic shifts, and unequal resource distributions (Muhammad et al. 2019). In this regard, the global smart city discussion serves as both a local innovation and an international collaboration platform.

Lastly, the technological aspects of smart cities, including the Internet of Things, infrastructure, artificial intelligence-powered service delivery, and renewable energy sources, present vital opportunities for modifying city governance and empowering citizens (Brzeziński and Wyrwicka 2022). It is imperative to understand the embeddedness, contestation, and adaptation of these technologies across various contexts to gain insight into their actual impact. The global picture of smart cities, therefore, serves as a framework for informed, inclusive, and ethically sound urban futures.

2.3 Purpose of the Study

Driven by the urgent need for a livable and sustainable future, this study examines the anatomy of smart cities from a global perspective to unravel the intricacies of their design, implementation, and impact. From its evolution, key components, challenges, case study, and future Trends and Directions, this study highlights the need for smart cities to address urban congestion in most cities worldwide.

2.4 Evolution of Urbanization and Technology in Relation to Smart Cities

The process of urbanisation is a diverse and comprehensive process that has shaped human settlement over the ages, beginning with early civilisations and extending to modern megacities. According to Henderson and Turner (2020), this entails the process of rural communities transitioning into urban hubs, a trend that can be traced back to ancient Mesopotamia during the Uruk Period. This initial urbanisation is associated with the expansion of more cities due to climate-resistant living and the leadership of the Roman Empire in urban planning and infrastructure systems across Europe, North Africa, and the Middle East (Attaran 2023).

The great turning point was the Industrial Revolution of the eighteenth and nineteenth centuries, during which innovations such as the steam engine, railways, and mass production led to the emergence of industrial cities (Kelly et al. 2021; Satam et al. 2023). The challenges that emerged with the increasing density and complexity of cities included, but were not limited to, traffic jams, environmental degradation, social disparity, and infrastructure overload (Shen et al. 2024). These developments collectively paved the way for the advent of the smart city paradigm in the mid-twentieth century (Keshavarzi et al. 2021). In addition, a five-layer conceptual roadmap, as presented in Fig. 2.1, outlines the systematic tracing of the trajectory to smart urbanism.

Layers in Fig. 2.1 reflect the functional interdependence and temporal evolution in the evolution of moving towards smart city frameworks. Most importantly, the last step involves the introduction of innovative technologies and energy systems. One example of this evolution is Smart Energy Regions (SER), which are localised, renewable-based energy ecosystems enabled by intelligent management tools. According to Atack et al. (2022), SERs enhance energy independence and sustainability, particularly about climate change and energy security. The very idea of the digital twin, a virtual, dynamic, and real-time model of real-world urban environments, is equally transformative. Satam et al. (2023) report that digital twins can support planning, predictive analytics, and the optimisation of operations through simulation-driven planning, enabling decision-makers to manage complex urban systems more effectively. All these developments serve to demonstrate that smart cities are not fixed entities but rather an ongoing evolution that requires the engagement of all stakeholders. Their actualisation is based on the integration of historical, technological, and governance frameworks, supported by a dedication to innovation, inclusivity, and sustainability.

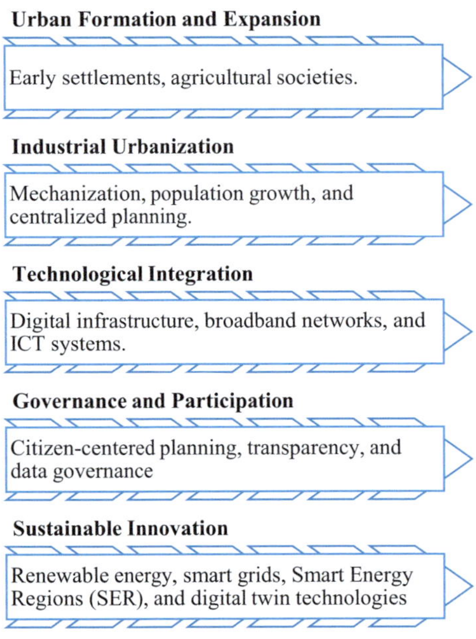

Fig. 2.1 Conceptual roadmap to smart cities. (Source: Author's Compilation)

2.5 Key Milestones in the Development of Smart Cities Global

Smart cities have emerged as a solution to address urbanisation challenges, such as resource depletion and climate change, by leveraging modern technology to create cities with improved living conditions in urban areas. Nevertheless, the first stages of smart cities can be regarded as a phenomenon that began in the 1990s, when cities like Singapore and Barcelona introduced innovative city concepts that integrated technology into urban infrastructure and services (Nižetić et al. 2020). For example, Singapore's Intelligent Island initiative of 1997 was to enhance governance, public services and economic growth. Barcelona adopted smart sensors and data analytics to monitor transportation, waste management and energy efficiency. Similarly, the Smart City initiative in Amsterdam, launched in 2008, focused on fostering dialogue among government, business, and citizens to develop innovative solutions for the city's issues.

The blend of the first two decades of the twenty-first century gave rise to mature, innovative city projects and initiatives, which were naturally driven by the recognition of technology and data-driven approaches as primary tools for urban development (Khan et al. 2022). Smart cities like Barcelona, Singapore, and Seoul lead the adoption of smart technologies and innovative urban solutions, deploying smart grids, sensor networks, and digital governance platforms to advance efficiency, sustainability, and livability (Alavi et al. 2018; Moura and de Abreu e Silva 2021).

In addition, the COVID-19 pandemic, through unleashing a complex web of challenges for cities worldwide, also increased the adoption rates of innovative city technologies and practices. As shown by Butt (2021), the digital solutions that cities developed in a short period to address the spread of the disease, adapt to changing circumstances, and maintain the continuity of vital services included contact tracing apps, remote work platforms, and telemedicine. Following Javed et al. (2022), the pandemic highlighted the significance of urban resilience, adaptability, and innovation for urban planners and city managers, motivating cities to adopt digital transformation and smart city strategies more than ever. Today, the expansion of smart cities worldwide is expected to develop in line with the latest trends in climate change, urbanisation, and digitalisation. For instance, cities will increasingly apply technologies such as artificial intelligence, blockchain, and 5G networks to address complex urban challenges, enhance service delivery, and improve the community's quality of life (Aurigi and Odendaal 2022). Furthermore, collaboration and knowledge formation among cities will be necessary for creating innovation, scaling up successful initiatives, and building better, more resilient, inclusive, and sustainable urban futures. From here, the chapter moves to show the various key components of smart cities.

2.6 Key Components of Smart Cities

Smart cities are reactive, intelligent, interconnected, and sustainable cities, comprising the following components.

2.6.1 Infrastructure

As Kasznar et al. (2021) have shown, infrastructure forms the backbone of smart cities, ensuring uninterrupted connectivity and smooth operations across sectors. Moreover, as one of the smart city components, infrastructure enables smart cities to advance urban lives with technological innovations and data-driven solutions, such as dynamic road networks, shared mobility services, electric vehicles, and autonomous vehicles (Chang et al. 2020). It also helps to relocate cities to greener environments by utilising renewable energy sources (Keshavarzi et al. 2021). Finally, the infrastructural components of smart cities also include communication networks, such as the broadband networks in Seoul and Tokyo, which enable residents to access high-speed internet and make data-driven decisions.

2.6.2 Governance and Policy

As pointed out by Khang et al. (2022), governance and policy frameworks are key to the success of smart cities, as they form the foundation for fruitful cooperation, innovation, and sustainable development. It enables the development of smart city features, such as regulatory frameworks, public-private partnerships (PPPs), and citizen engagement, to be successfully implemented. For instance, regulations, which are rules or directives established and enforced by an authority, may differ, with technology and infrastructure falling under the purview of zoning ordinances, construction codes, data privacy laws, and cybersecurity measures (Khalimon et al. 2020). Moreover, PPPs, which combine the strengths of the government and private sectors, allow governments to accelerate project delivery, decrease risk, and control costs (Kirimtat et al. 2020). Ultimately, citizen engagement enables citizens to hold the government accountable, collectively develop solutions, and participate in the decision-making process (Browne 2020).

2.6.3 Technology Integration

Technology integration is undeniably a critical factor in smart cities. As pointed out by Tura and Ojanen (2022), it allows for the smooth combination of digital technology to better urban infrastructure, services, and people's lives. Some of those factors include the Internet of Things, big data analytics, and AI. IoT, for instance, is a pillar of smart city infrastructure, enabling physical devices to communicate with the internet in real-time for two-way data transmission. As Bauer et al. (2021) reveals, IoT devices, such as smart meters and traffic sensors, can reduce energy waste and improve road safety. Moreover, through big data analytics, machine learning, and predictive models, smart cities can also discover traffic patterns and identify where congestion is most likely to occur (Luckey et al. 2021). Ultimately, AI enhances the efficiency and innovation of smart city processes while also facilitating informed decision-making and upgrading city services. For instance, Katz (2021) demonstrated that AI-enhanced chatbots and virtual assistants facilitate and automate public transportation and waste management services, ensuring the public is well-served.

2.6.4 Sustainability and Resilience

As one of the key components of Smart Cities, sustainability and resilience, as illustrated in Fig. 2.2, are at the core of innovative city development, reflecting efforts to create environmentally responsible cities that can adapt to a changing environment to serve both present and future generations (Zeng et al. 2022).

As shown in Fig. 2.2, Sustainability and Resilience, as key components of smart cities, focus on lowering emissions, reducing pollution, and conserving natural resources through effective environmental impact management. Specifically, Information-based methods and technologies are used to monitor and mitigate the environmental impacts of transportation, energy, waste management, and urban planning (Ramirez Lopez and Grijalba Castro 2020). Predictive analytics and machine learning algorithms facilitate

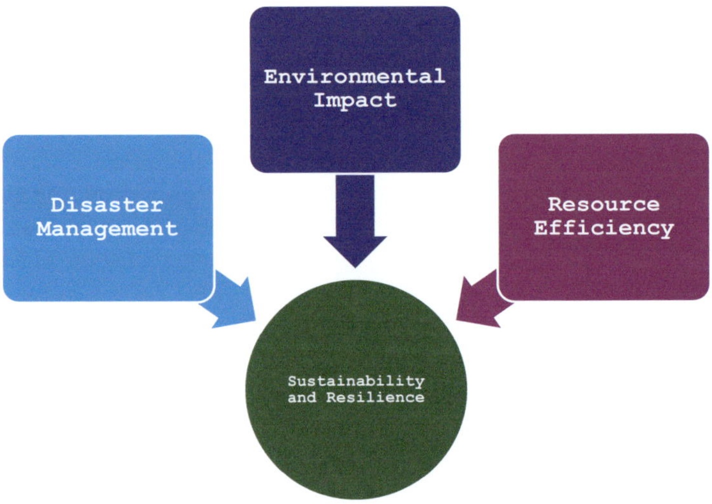

Fig. 2.2 Sustainability and resilience in smart cecities. (Source: Author's Compilation)

the forecasting of extreme weather events and the development of evacuation plans, which are essential for disaster management and resilience planning. Resource efficiency, as another component, is concerned with innovative grid technologies which allow utilities to efficiently manage their energy demand and integrate renewable energy sources (Appio et al. 2019).

2.7 Case Studies: Global Perspective on Smart City Initiatives and National Strategies

Smart cities around the world have already taken significant steps to integrate new technologies that promote sustainability, energy savings, and reduced carbon footprints. According to Sharifi et al. (2021), the emergence of smart initiatives is preconditioned by the necessity of more innovative and resilient urban environments. Although the primary focus is usually paid to separate projects in the cities, it is also necessary to focus on national-level policies and strategic plans that inform the political and institutional environment of the specified developments. Most national governments have developed comprehensive smart city policies, digital transformation plans, and funding programs that guide and support metropolitan innovation. The following are both local initiatives in cities and more general national policies in various parts of the world.

2.7.1 North America

New York City in North America has become a leading example of applying intelligent technologies to spur economic development and increase urban efficiency. The LED lighting retrofit and an Automated Meter Reading system are initiatives that can be considered to achieve sustainability objectives, with more than $800,000 in yearly energy and wastewater savings (Hussain 2024). These developments can be seen within the scope of national policies, such as the U.S. Smart Cities Initiative, introduced by the White House in 2015, which aims to fund and provide technical assistance to more than 70 cities for their energy, transportation, and public safety projects.

Likewise, the federal government of Canada initiated the 2017 Smart Cities Challenge, in which cities were incentivised to develop innovative city solutions. In this context, Toronto established the Smart Cities Working Group (SCWG) and advanced inclusive, climate-resilient, and mobility-oriented urban development (Israilidis et al. 2021; Shamsuzzoha et al. 2021).

2.7.2 Europe

European cities also benefit from the presence of unified regional plans, including the Digital Agenda for Europe and the Horizon 2020 initiatives, which the European Union funds to facilitate funding and regulatory coordination. Amsterdam, in particular, has implemented more than 170 innovative city initiatives since 2009, including floating housing projects and the use of solar energy (Kutty et al. 2020). They are supported by the Dutch Digital Government Strategy, which encourages open data, citizen participation, and digital inclusion.

Barcelona, which has frequently been regarded as a model of a smart city, has tailored its projects to the EU Horizon 2020 framework. Its City OS and the Smart Citizen platforms combine data from various areas to improve governance and civic engagement (Heidari et al. 2022). The national government of Spain is also enhancing innovative city development through its Digital Spain 2025 plan, focusing on cooperation between the state and businesses, as well as digital infrastructure.

2.7.3 Asia

In Asia, an example of the top-down approach is Singapore's Smart Nation program, established in 2014, which covers the entire nation. Financial allocations have facilitated the smooth integration of AI, IoT, and digital governance systems within industries (Reis et al. 2022; Almalki et al. 2023). This centralised approach demonstrates how powerful national policy can facilitate the rapid adoption of innovative technologies.

At the national level, South Korea aims to promote digital innovation, which is comprised of projects such as Songdo and Seoul, that utilise real-time data analytics and AI to enhance traffic management, public safety, and environmental surveillance (Brzeziński and Wyrwicka 2022; Alagirisamy and Ramesh 2022). These initiatives align with the Digital New Deal in South Korea, a national stimulus package that prioritises smart infrastructure and green innovation.

2.7.4 Africa

In Africa, although many smart city activities remain nascent, national policies are playing an increasingly significant role. In Kenya, the state has promoted smart infrastructure through its Vision 2030 plan, which includes digital economy ambitions. This broader ambition is evident in the use of ICT in Nairobi to support public Wi-Fi and e-governance (Wang et al. 2019; Alagirisamy and Ramesh 2022).

South Africa has initiated the National Integrated ICT Policy White Paper to facilitate a digitally inclusive society. Cape Town benefits from this by incorporating real-time IoT data into the management of utilities and urban services (Nižetić et al. 2020). Nevertheless, they are frequently disrupted by systemic deficiencies in financing, physical infrastructure, and political organisation.

2.7.5 Oceania

Both Australia and New Zealand have established national initiatives in favour of smart urbanisation. In 2016, the Australian Smart Cities Plan was published, establishing a federal roadmap for investing in digital infrastructure and driving urban innovation. In this context, Sydney has implemented fibre-optic systems and promoted e-mobility (Aurigi and Odendaal 2022).

In New Zealand, the Digital Government Strategy has prioritised data-driven governance, contributing to initiatives such as Safe Swim in Auckland and Farms for the Future, as well as the application of real-time analytics to deliver better public services and related environmental surveillance capabilities (Sancino and Hudson 2020).

2.8 Challenges and Opportunities

The concept of a smart city has evolved from a utopian vision to a practical goal for urban planners and governments globally in the twenty-first century. Nonetheless, the path to becoming a smart city is fraught with obstacles and opportunities. Among these obstacles are:

2.8.1 Data Privacy and Security Concerns

Although smart cities appear to hold great promise, privacy and data security issues exist. As noted by Kozłowski and Suwar (2021), many systems are interconnected, and the functioning of smart cities necessitates extensive data collecting, which might result in privacy and personal information breaches. Furthermore, the fact that sensor data may inevitably compromise people's integrity raises ethical questions as well (Rao and Deebak 2023). IoT-based smart city solutions have also brought up security issues, such as cyber and physical security flaws in smart grids (Fabrègue and Bogoni 2023). Finally, system failures may raise serious security issues, jeopardising privacy, as wireless sensor network (WSN) security is crucial for smart cities (Ahmed 2020).

2.8.2 Digital Divide and Equity Issues

The digital divide, referring to the gap between those with access to digital technologies and those without, is one of the most crucial obstacles in smart city development. Smart city technologies offer efficient solutions, sustainability, and efficiency, but adoption and access still need to be distributed more evenly among populations, resulting in limited opportunities, information, and services. Nižetić et al. (2020) showed that this situation makes it challenging to deploy intelligent cities, as most programs are centred in affluent areas, leaving underprivileged people behind. According to Shen et al. (2024), the unequal distribution of intelligent services negatively impacts the quality of life. Toli and Murtagh (2020) also demonstrated that the uneven distribution of intelligent services negatively impacts people's quality of life and their ability to access essential services, such as healthcare and education. As such, a multifaceted strategy, including infrastructure investments, legislative interventions, and community participation, is necessary to address this issue.

2.8.3 Scalability and Interoperability

As O'Dwyer et al. (2019) puts it, scalability guarantees a smooth user experience, drives adoption, and keeps the ecosystem decentralised, while interoperability expands the utility of smart cities, facilitates collaboration with other networks, and unlocks new use cases. As such, since smart cities utilise many IoT devices, which are primarily designed to be cost-effective and resource-efficient, they are often characterised by limited processing power, memory, and battery life (Katz 2021). In the words of Bhushan et al. (2020), these issues hinder the development of deep interoperability norms and high-level security protocols, which typically require significantly more effort.

2.8.4 Financial Constraints and Funding Models

The introduction of smart cities can suffer from financial roadblocks such as lack of public investment capacity or unavailability of private financing due to identified risks, long maturity times and even the impossibility of monetizing the smart solution (Butt 2021). As such, financing suitable technologies to meet urban needs can become a significant challenge for most cities, as funds for these technologies and services require innovative financing options, such as capital management and business models (Ramirez Lopez and Grijalba Castro 2020).

2.8.5 Opportunities for Collaboration and Knowledge Sharing

Collaboration is a common obstacle during smart city development, often due to internal silos, lack of structure, and an uncoordinated vision, which may delay project implementation. On the other hand, cooperation and knowledge exchange create opportunities for cities to address complex urban problems more successfully (Javed et al. 2022). Thus, close cooperation will equip cities with modern research, experience, and innovation centres that will foster innovation and experimentation.

2.9 Future Trends and Directions

Addressing the ever-increasing requirements of inhabitants has been shown to settle with smart city development. As Alagirisamy and Ramesh (2022) pointed out, although digital transformation is key to success, data-driven decision-making is crucial. As a result, these patterns and recommendations represent the future of smart cities.

2.9.1 Emergence of Smart City Ecosystems

Smart city ecosystems are an emerging global trend in which stakeholders form partnerships to create interconnected, efficient, and green urban centres. However, according to Wang et al.

(2019), these ecosystems involve government agencies, technology companies, startups, academia, civil society organisations, and citizens collaborating to solve complex urban problems and capitalise on opportunities for innovation and growth. The fact that only some entities address these challenges alone leads to the appearance of intelligent city ecosystems (Huang et al. 2021). Consequently, public-private partnerships, cross-sectoral collaborations, and citizen engagement initiatives are the motor behind innovative city developments around the globe. Technological innovations and big data analytics allow cities to develop more integrated and interoperable city networks where various services and systems can share information without delay (Brzeziński and Wyrwicka 2022).

Similarly, innovation neighbourhoods and urban innovation hubs also lead the way for smart city communities. As Tanwar et al. (2022) noted, these hubs serve as a melting pot of technology companies, startups, research institutions, and government agencies that collaborate to create and test new solutions for urban problems, thereby stimulating economic development, job creation, and social innovation. For example, cities such as Barcelona, Singapore, and Amsterdam have developed cutting-edge innovation ecosystems to test new technologies and urban governance models (Hussain 2024).

2.9.2 Integration of Emerging Technologies (e.g., 5G, Blockchain)

The introduction of 5G and blockchain in the realm of smart city development is a significant development. However, it's important to move beyond a simplistic view of these technologies as inherently beneficial. Understanding their operational complexities and socio-political implications is crucial. For example, 5G isn't just about faster mobile communication. It serves as a structural enabler of ultra-reliable low-latency communications (URLLC), massive machine-type communications (MTC), and enhanced mobile broadband (eMBB) (Shaik et al. 2024). These capabilities are crucial for real-time applications, such as autonomous vehicles, remote surgeries, smart grid control, and smart traffic systems (Almalki et al. 2023).

Similarly, blockchain, when applied beyond cryptocurrencies, provides encrypted, decentralised data services that are essential for citizen-friendly services such as identity verification, participatory governance, or peer-to-peer energy trading (Reis et al. 2022). When these technologies are combined, they create highly efficient, secure, and interoperable systems. Notably, Nagargoje et al. (2025) have shown that blockchain can secure 5G-based IoT data transfers, ensuring integrity and audibility, as evidenced in sectors such as logistics and healthcare.

Despite the potential, there are significant hurdles to overcome. Critics argue that blockchain's energy-intensive consensus mechanism may conflict with sustainability goals, and the infrastructure demands of 5G could limit its use in underdeveloped areas (Rajavat et al. 2024). Moreover, the absence of consistent regulatory frameworks creates legal uncertainty, particularly in areas such as data management and interoperability. As proposed by Heidari et al. (2022), the implementation of these technologies necessitates a clear ethical vision and policy consideration. This is essential to realize the vision of a truly inclusive and resilient urban ecosystem. The success of this integration will depend not only on technological performance but also on the institutional and societal contexts in which it is deployed.

2.9.3 Impact of Global Events (e.g., Climate Change, Pandemics)

Climate change is one of the most pressing and critical challenges of our time; as average temperatures continue to rise year after year, extreme weather events become increasingly common, and sea levels continue to rise. This trend has been shown to endanger lives and prosperity in cities (Sharifi et al. 2021). Nastjuk et al. (2022) argue that smart cities can combat climate change by leveraging technology, analytics, and innovation to reduce greenhouse gas emissions, enhance resilience, and promote sustainability. The implementation of green infrastructure, renewable energy installations, and sustainable transportation systems can reduce carbon footprints, improve air quality, and mitigate the urban heat island effect (Javed et al. 2022). Pandemic events like COVID-19 underscore the need for robust urban systems. Smart cities have facilitated the response to the pandemic through the use of technology and data, enabling the tracking of virus spread, supporting remote work and learning, and the provision of essential services (Butt 2021). As a result, the impact of global events on intelligent cities is likely to drive increased financing and innovation in the fields of resilience, sustainability, and public health (Katz 2021).

2.9.4 Citizen-Centric Approaches to Smart City Development

The rise of citizen-centric intelligent city development is a rapidly emerging trend that emphasises the needs, demands, and expectations of city residents. Implementing this method aims to enhance the quality of life, social inclusion, and active participation of residents in the development of their cities (Khan et al. 2022). Transparency, accountability and participation also form a core part of good governance. Therefore, cities should make the information, data, and decision-making processes that concern residents available to them through active participation in the democratic process and the authorities' accountability. This approach fosters trust with everyone and provides a solid foundation for collaboration and partnership in innovative city development. Equity and inclusion are fundamental, ensuring that all city residents, regardless of their socioeconomic status, age, or disability, benefit from innovation (O'Dwyer et al. 2019). By doing this, cities must eliminate digital gaps, access problems, and social inequities to ensure the levelling of digital transformation in urban areas (Bhushan et al. 2020).

2.10 Conclusion

This study examined the Anatomy of smart cities from a Global Perspective. The study found that the emergence of smart cities evolved from the pattern of urbanisation and technological progression. Additionally, the research showed that four elements of Smart Cities—infrastructure, governance, policy frameworks, technology integration, and sustainability—revolve around achieving a connected, equitable, and resilient urban system. However, although smart city initiatives have several positive aspects, issues such as data privacy concerns, digital divides, scalability problems, financial challenges, and interoperability issues also arise. Nevertheless, future trends, such as innovative city ecosystems and 5G integration, were shown to pave the way for smart cities to blossom.

References

Ahmed S (2020) Security and privacy in smart cities: challenges and opportunities. Int J Eng Trends Technol 68(2):1–8

Alagirisamy B, Ramesh P (2022) Smart sustainable cities: principles and future trends. In: Sustainable cities and resilience: select proceedings of VCDRR 2021. Springer Singapore, pp 301–316

Alavi AH, Jiao P, Buttlar WG, Lajnef N (2018) Internet of Things-enabled smart cities: State-of-the-art and future trends. Measurement 129:589–606

Almalki FA, Alsamhi SH, Sahal R, Hassan J, Hawbani A, Rajput NS, Saif A, Morgan J, Breslin J (2023) Green IoT for eco-friendly and sustainable smart cities: future directions and opportunities. Mob Netw Appl 28(1):178–202

Appio FP, Lima M, Paroutis S (2019) Understanding Smart Cities: innovation ecosystems, technological advancements, and societal challenges. Technol Forecast Soc Chang 142:1–14

Atack J, Margo RA, Rhode PW (2022) Industrialization and urbanization in nineteenth century America. Reg Sci Urban Econ 94:103678

Attaran M (2023) The impact of 5G on the evolution of intelligent automation and industry digitization. J Ambient Intell Humaniz Comput 14(5):5977–5993

Aurigi A, Odendaal N (2022) From "smart in the box" to "smart in the city": rethinking the socially sustainable smart city in context. In: Sustainable smart city transitions. Routledge, pp 53–68

Bauer M, Sanchez L, Song J (2021) IoT-enabled smart cities: evolution and outlook. Sensors 21(13):4511

Bhushan B, Khamparia A, Sagayam KM, Sharma SK, Ahad MA, Debnath NC (2020) Blockchain for smart cities: a review of architecture, integration trends and future research directions. Sustain Cities Soc 61:102360

Browne NJ (2020) Regarding smart cities in China, the north and emerging economies—one size does not fit all. Smart Cities 3(2):11

Brzeziński Ł, Wyrwicka MK (2022) Fundamental directions of the development of the smart cities concept and solutions in Poland. Energies 15(21):8213

Butt TA (2021) Future smart cities: vision, challenges and technology trends. In: Proceedings of the 2021 9th international conference on information technology: IoT and smart city, pp 353–359

Chang V, Sharma S, Li CS (2020) Smart cities in the 21st century. Technol Forecast Soc Chang 153:119447

Chen Z, Gan W, Wu J, Lin H, Chen CM (2024) Metaverse for smart cities: a survey. Internet Things Cyber-Phys Syst 4:203–216

Fabrègue BF, Bogoni A (2023) Privacy and security concerns in the smart city. Smart Cities 6(1):586–613

Heidari A, Navimipour NJ, Unal M (2022) Applications of ML/DL in the management of smart cities and societies based on new trends in information technologies: a systematic literature review. Sustain Cities Soc 85:104089

Henderson JV, Turner MA (2020) Urbanization in the developing world: too early or too slow? J Econ Perspect 34(3):150–173

Huang H, Yao XA, Krisp JM, Jiang B (2021) Analytics of location-based big data for smart cities: opportunities, challenges, and future directions. Comput Environ Urban Syst 90:101712

Huda NU, Ahmed I, Adnan M, Ali M, Naeem F (2024) Experts and intelligent systems for smart homes' Transformation to Sustainable Smart Cities: a comprehensive review. Expert Syst Appl 238:122380

Hussain I (2024) Secure, sustainable smart cities and the internet of things: perspectives, challenges, and future directions. Sustainability 16(4):1390

Israilidis J, Odusanya K, Mazhar MU (2021) Exploring knowledge management perspectives in smart city research: a review and future research agenda. Int J Inf Manag 56:101989

Javed AR, Shahzad F, Ur Rehman S, Zikria YB, Razzak I, Jalil Z, Xu G (2022) Future smart cities: requirements, emerging technologies, applications, challenges, and future aspects. Cities 129:103794

Kasznar APP, Hammad AW, Najjar M, Linhares Qualharini E, Figueiredo K, Soares CAP, Haddad AN (2021) Multiple dimensions of smart cities' infrastructure: a review. Buildings 11(2):73

Katz C (2021) Splanetary urbanization. Int J Urban Reg Res 45(4):597–611

Kelly B, Papanikolaou D, Seru A, Taddy M (2021) Measuring technological innovation over the long run. Am Econ Rev Insights 3(3):303–320

Keshavarzi G, Yildirim Y, Arefi M (2021) Does scale matter? An overview of the "smart cities" literature. Sustain Cities Soc 74:103151

Khalimon EA, Vikhodtseva EA, Obradović V (2020) Smart cities today and tomorrow–world experience. In: Institute of Scientific Communications Conference. Springer International Publishing, Cham, pp 1340–1347

Khan A, Aslam S, Aurangzeb K, Alhussein M, Javaid N (2022) Multiscale modeling in smart cities: a survey on applications, current trends, and challenges. Sustain Cities Soc 78:103517

Khang A, Hahanov V, Litvinova E, Chumachenko S, Hajimahmud VA, Alyar AV (2022) The key assistant of smart city: sensors and tools. In: AI-centric smart city ecosystems. CRC Press, pp 271–280

Kirimtat A, Krejcar O, Kertesz A, Tasgetiren MF (2020) Future trends and current state of smart city concepts: a survey. IEEE Access 8:86448–86467

Kozłowski W, Suwar K (2021) Smart city: definitions, dimensions, and initiatives

Kutty AA, Abdella GM, Kucukvar M, Onat NC, Bulu M (2020) A system thinking approach for harmonizing smart and sustainable city initiatives with United Nations sustainable development goals. Sustain Dev 28(5):1347–1365

Liang J, Pu Y, Chen J, Liu M, Ouyang B, Jin Z, Ge W, Wu Z, Yang X, Qin C, Wang C (2025) Global prevalence, trend and projection of myopia in children and adolescents from 1990 to 2050: a comprehensive systematic review and meta-analysis. Br J Ophthalmol 109(3):362–371

Luckey D, Fritz H, Legatiuk D, Dragos K, Smarsly K (2021) Artificial intelligence techniques for smart city applications. In: Proceedings of the 18th international conference on computing in civil and building engineering: ICCCBE 2020. Springer International Publishing, pp 3–15

Moura F, de Abreu e Silva J (2021) Smart cities: definitions, evolution of the concept, and examples of initiatives. In: Industry, innovation and infrastructure. Springer International Publishing, Cham, pp 989–997

Muhammad K, Lloret J, Baik SW (2019) Intelligent and energy-efficient data prioritization in green smart cities: current challenges and future directions. IEEE Commun Mag 57(2):60–65

Nagargoje Y, Shravan A, Ambudkar B (2025) Blockchain-enhanced 5G networks: securing IoTDriven medical supply chains. In: 2025 1st international conference on aiml-applications for engineering & technology (ICAET). IEEE, pp 1–6

Nastjuk I, Trang S, Papageorgiou EI (2022) Smart cities and smart governance models for future cities: current research and future directions. Electron Mark 32(4):1917–1924

Nižetić S, Šolić P, Gonzalez-De DLDI, Patrono L (2020) Internet of Things (IoT): opportunities, issues and challenges towards a smart and sustainable future. J Clean Prod 274:122877

O'Dwyer E, Pan I, Acha S, Shah N (2019) Smart energy systems for sustainable smart cities: current developments, trends and future directions. Appl Energy 237:581–597

Rajavat A, Bhardwaj V, Kaur N, Rawat R, Rawat A, Jadon GS (2024) Sustainable futures: navigating blockchain's energy dilemma. In: Online social networks in business frameworks. Wiley, pp 85–112

Ramirez Lopez LJ, Grijalba Castro AI (2020) Sustainability and resilience in smart city planning: a review. Sustainability 13(1):181

Rao PM, Deebak BD (2023) Security and privacy issues in smart cities/industries: technologies, applications, and challenges. J Ambient Intell Humaniz Comput 14(8):10517–10553

Reis J, Marques PA, Marques PC (2022) Where are smart cities heading? A meta-review and guidelines for future research. Appl Sci 12(16):8328

Sancino A, Hudson L (2020) Leadership in, of, and for smart cities–case studies from Europe, America, and Australia. Public Manag Rev 22(5):701–725

Satam H, Joshi K, Mangrolia U, Waghoo S, Zaidi G, Rawool S, Thakare RP, Banday S, Mishra AK, Das G, Malonia SK (2023) Next-generation sequencing technology: current trends and advancements. Biology 12(7):997

Shaik RB, Nagaradjane P, Ioannou I, Sittakul V, Vasiliou V, Pitsillides A (2024) AI/ML-aided capacity maximization strategies for URLLC in 5G/6G wireless systems: a survey. Comput Netw 249:110506

Shamsuzzoha A, Nieminen J, Piya S, Rutledge K (2021) Smart city for sustainable environment: a comparison of participatory strategies from Helsinki, Singapore and London. Cities 114:103194

Sharifi A, Allam Z, Feizizadeh B, Ghamari H (2021) Three decades of research on smart cities: mapping knowledge structure and trends. Sustainability 13(13):7140

Shen J, Wang S, Wang Y (2024) Environmental inequality in peri-urban areas: a case study of Huangpu District, Guangzhou City. Land 13(5):703

Tanwar S, Popat A, Bhattacharya P, Gupta R, Kumar N (2022) A taxonomy of energy optimization techniques for smart cities: architecture and future directions. Expert Syst 39(5):e12703

Toli AM, Murtagh N (2020) The concept of sustainability in smart city definitions. Front Built Env 6:77

Tura N, Ojanen V (2022) Sustainability-oriented innovations in smart cities: a systematic review and emerging themes. Cities 126:103716

Wang J, Wang Y, Zhang D, Lv Q, Chen C (2019) Crowd-powered sensing and actuation in smart cities: current issues and future directions. IEEE Wirel Commun 26(2):86–92

Zeng X, Yu Y, Yang S, Lv Y, Sarker MNI (2022) Urban resilience for urban sustainability: concepts, dimensions, and perspectives. Sustainability 14(5):2481

3 The Role of Governance in Smart Cities' Evolution

Abstract

Smart cities governance, which utilises data-driven strategies, is crucial for enhancing decision-making and public services in cities; yet, its challenges remain unclear, necessitating a critical examination. Thus, this chapter examines the structural dimensions of smart city governance. It critiques the lack of universally accepted governance frameworks and examines the tension between technology-driven and people-centric approaches. Moreover, structural elements such as policy frameworks, regulatory mechanisms, decision-making processes, and stakeholder collaboration were discussed, with particular attention paid to the growing use of real-time data and its implications for privacy and democratic participation. The chapter also examined governance models, including centralised, decentralised, and hybrid, which were compared alongside their respective benefits and limitations.

Furthermore, the chapter's outcome warns against corporate dominance in governance and the superficiality of public participation. It also addressed broader issues such as digital divides, financial constraints, and fragmented regulatory practices. Through case studies of Singapore, Barcelona, and Copenhagen, the chapter illustrated the potential and limitations of smart city governance. Ultimately, this chapter calls for adaptive, participatory, and ethically grounded governance models that prioritise inclusivity, accountability, and sustainability over techno-utopian visions of urban progress.

3.1 Introduction

The concept of the smart city has been the subject of increasing research and urban policy discussions over the last 20 years. Although the term has gained mainstream status, it is conceptually vague, with many classifications and overlaps with adjacent concepts such as digital cities, sustainable urbanism, and e-governance (Lytras and Şerban 2020). In essence, the term smart cities involve the incorporation of information and communication technologies (ICTs), which, in turn, improve the quality and efficiency of urban services, enhance citizen well-being, and promote sustainable economic growth (Humayun et al. 2020; Abdel-Razek 2021). The most common areas are smart governance, smart mobility, smart living, and smart infrastructure (Kuzior et al. 2023). Nevertheless, according to Lim et al. (2023), the subsequent development of interpretation since the 1990s resulted in fragmented interpretations due to various regional interests and technological capacities.

Technological aspects, specifically the use of sensor-equipped and wireless-connected ICTs,

V. Yorucu et al., *Modern Approach of Resilient and Sustainable Smart Cities*, Sustainable Development Goals Series, https://doi.org/10.1007/978-3-032-05715-0_3

have become a common thread in the smart city narrative (Noori et al. 2020; Razmjoo et al. 2022). These technologies combine to allow real-time optimisation of urban systems, including transport, healthcare, public safety, and utilities. However, this techno-centric agenda has garnered criticism, suggesting that it may undermine social justice and citizen empowerment (Oliveira et al. 2020; Estevez et al. 2021). The response to this has been the rise of a human-centred approach embodied in the concept of citizens as co-creators and active participants as opposed to passive receivers of services. Nevertheless, despite these conceptual innovations, conceptual barriers to practical integration remain, such as unequal access to digital infrastructure, elevated maintenance costs, and concerns regarding data privacy and algorithmic control (Ooms et al. 2020; Kutty et al. 2020). As such, governance becomes more vital as smart cities become intertwined. Studies show that proper governance models are necessary to regulate the relationship between stakeholders in the state and the corporate sector, to impose ethical principles in data usage, and to promote inclusive engagement (De Guimarães et al. 2020; Nastjuk et al. 2022). Qin and Qi (2021) suggested that governance in smart cities should be responsive and adaptive, underpinned by strong legal systems and institutional capacity.

Moreover, the leading role of governance in the development of the smart city cannot be overemphasised. For instance, Gracias et al. (2023) demonstrated that transparent and accountable governance mechanisms are necessary to foster innovation and collaboration across sectors, thereby strengthening the resilience of urban infrastructure. As such, this chapter provides a clearer understanding of the role of governance in the development of smart cities. This is done through intellectual extrapolation constructed by the author, who has integrated the theoretical and practical frameworks presented in the existing literature and the most prominent global case studies. The multi-stakeholder approach employed, based on real-life governance models such as those in Barcelona and Singapore, is not a universal solution but rather a proposed analytical tool for understanding partnerships among various actors in smart city strategies. It critically scrutinises interactions and provides a theoretical perspective for assessing governance mechanisms in diverse urban settings.

Finally, this chapter uses collaborative governance theory as an analytical guide. Ooms et al. (2020), define collaborative governance as processes and arrangements that, in a broad sense, involve the coordination and joint decision-making between various public agencies, stakeholders in the private sector, and civil society. With the help of this theoretical lens, the chapter critically examines the enabling conditions and constraints provided by various models of governance on cross-sector collaboration, including how the role of stakeholders is negotiated and how tensions between centralisation and decentralisation are practically addressed.

3.2 Overview of Governance Structures

Cities have been regarded as a key factor in shaping the world's future evolution. This is because many people live in cities, and the high population density leads to problems such as traffic congestion, waste management issues, and air pollution (Nastjuk et al. 2022). Many cities have participated in and utilised projects that involve and use the latest advances in information and communication technology (ICT) (De Guimarães et al. 2020). Hence, the increasing importance of information technology and human resource strategies in metropolitan areas is primarily attributed to the development of Smart City Governance (SCG) (De Guimarães et al. 2020). Oliveira et al. (2020) argued that these elements help governments achieve their goals and enhance the performance and productivity of governance. Nastjuk et al. (2022) notes that SCG advocates for the participation and involvement of various actors in urban environments. It is intended to encourage citizen participation in governance processes and invite players from all walks of life. Kirimtat et al. (2020) further explain the significance of stakeholder engagement in SCG, as it facilitates the development, implementation, and evaluation of strategies for a smooth and smart city.

The complexity and nature of agencies and organisations engaged in delivering smart city capabilities in metropolitan regions are mirrored in the governance structures of smart cities. Grossi et al. (2020) point out that the trend of increased urbanisation in the global population has raised the demand for smart city governance to ensure better investments, laws and regulations, innovation, inclusion, and efficiency. Similarly, Masik et al. (2021) argued that proper engagement with multiple stakeholders reinforces partnerships and develops tailored approaches for each city. For example, Barcelona earned its reputation as one of the world's smartest cities thanks to this collaborative governance approach (Kutty et al. 2020).

In addition, decision-making in cities is increasingly focusing on smart city governance systems for effectiveness and accountability, driven by real-time data generated from IoT devices and platforms, which include parameters such as air quality, traffic, energy generation and consumption, as well as citizen participation (Kuzior et al. 2023). Lim et al. (2023) asserts that utilising this data enables municipal officials to discover new patterns, develop a better understanding of urban developments and make policies that align with the actual situation. Again, data-driven governance frameworks in smart cities promote resilience, sustainability, and collaboration by reducing carbon emissions, supporting renewable energy sources, and increasing green areas in urban planning regulations (Dana et al. 2022). Furthermore, resilient cities aim to mitigate the adverse effects of natural disasters, economic recessions, and climate change (Tan and Taeihagh 2020). Finally, Smart city governance frameworks emphasise the importance of equality and diversity in cities, as socioeconomic disparities and digital divides can perpetuate inequality by denying certain groups access to opportunities and basic services (Xiao and Xie 2021). Collecetively, the theory of collaborative governance helps illustrate how different organizational forms enable or inhibit cooperation among stakeholders in smart city projects in this chapter.

3.3 Key Components of Smart Cities Governance

Smart city governance comprises various components that influence collaboration across multiple stakeholders, as illustrated in Fig. 3.1.

3.3.1 Policy Frameworks

Figure 3.1 illustrates that policy frameworks are a key component of smart city governance, serving as the main strategy and regulation for managing complex urban issues and urban development. Estevez et al. (2021) demonstrates that these frameworks are crucial for cities to champion equitable development, protect the public interest, and collaborate towards larger objectives. Moreover, Fig. 3.2 shows the policy component structure.

Figure 3.2 illustrates that smart city projects are developed within the defined vision and key objectives in the areas of ecological sustainability, economic prosperity, social participation, and

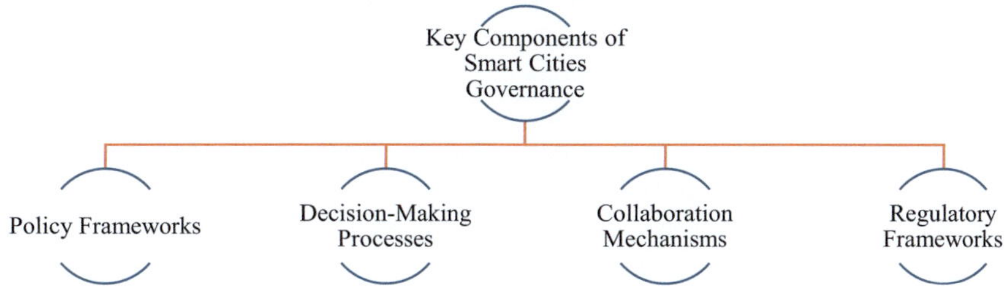

Fig. 3.1 Components of smart cities governance. (Source: Author's compilation)

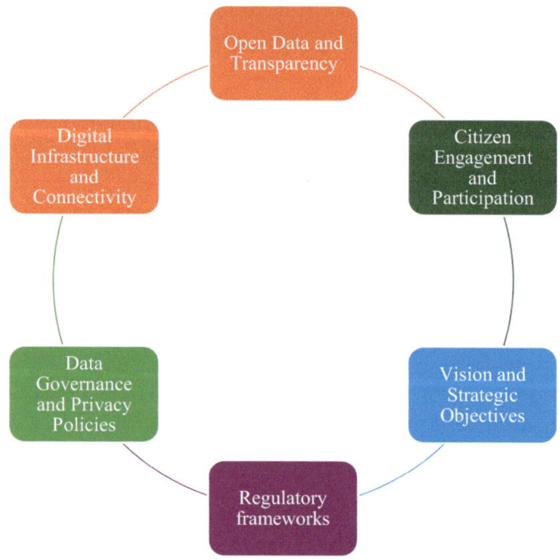

Fig. 3.2 Policy frameworks in smart city governance. (Source: Authors construction)

sustainability. For instance, regulatory frameworks and data governance policies dictate how smart city technologies will be utilised in terms of privacy, security, environmental safety and protection, as well as public health and safety (Noori et al. 2020). Additionally, policy frameworks encompass policies aimed at enhancing and sustaining digital infrastructure for smart cities, promoting inclusion, fostering sharing economies, and driving growth (Razmjoo et al. 2022). These frameworks connect sustainability with governance to achieve more desirable and sustainable cities.

3.3.2 Decision-Making Processes

As one of the components of smart city governance, decision-making Processes under smart cities refer to the processes of governance that entail identifying, evaluating, and implementing smart city initiatives to improve the city's quality of life (Lytras and Şerban 2020).

Figure 3.3 illustrates that problem identification is one of the most important processes, incorporating data analysis and needs identification (Abdel-Razek 2021), as well as stakeholders' engagement to recognise urban issues, opportunities, and aspirations (Humayun et al. 2020). It focuses on developing visions for sustainable development, economic growth, and social equity, aiming for inclusive and responsive outcomes.

3.3.3 Cross-Sector Collaboration in Practice

According to De Guimarães et al. (2020), collaboration is a priority for smart city governance, addressing complex problems through mechanisms such as multi-stakeholder partnerships, intergovernmental work, and public-private partnerships (PPPs) that leverage partners' capabilities to address jurisdictional issues. Ooms et al. (2020) noted that cross-sectoral working groups are dedicated to specific issues such as transportation, energy, public health, or social inclusion. The avenue for community engagement allows citizens to be actively involved in the decision-making process and contribute to social integration and trust in the government (Polese et al. 2021).

3.3.4 Regulatory Frameworks

Regulatory frameworks are a crucial aspect of smart city governance, shaping and providing the legal and institutional processes that control the

Fig. 3.3 Stakeholders engagement. (Source: Author's compilation)

creation, implementation, and operation of smart technology and initiatives (De Guimarães et al. 2020). This component encompasses data privacy rules that govern the collection, retention, processing, and dissemination of personal information by smart city technologies and platforms. These policies establish boundaries to protect personal privacy and ensure ethical data practices (Ooms et al. 2020). Some of the environmental issues covered by the regulations are energy efficiency regulations, waste management regulations, and green building standards, which aim to promote sustainability and reduce carbon emissions.

3.4 Governance Models in Smart Cities

Governance models in Smart Cities have emerged as a key differentiator in the style with which decisions are made, how projects are conducted, and the mode of service delivery (Abdel-Razek 2021). Logically, this part examines some of these models and how each deals with and controls cities.

3.4.1 Centralised Governance Models

Centralised governance is an organisational model in which data collection and control are centralised within a single organisation (Nastjuk et al. 2022). Centralised governance models in smart cities refer to the centralised decision-making process during the planning, implementation, and management of technological smart cities (Ullah et al. 2021). This maintains a hierarchical structure and typically involves using sophisticated technologies that enable centralised decision-making processes and enforcing uniform policies for rapid resource allocation and the implementation of smart city projects. However, the centralised model also has its difficulties—lack of responsibility and transparency, the possibility of forgetting about local community interests in favour of big city goals, and most importantly, dependency on the decision-making body that cannot process and react to the large amount of data generated by a smart city in time (Majeed et al. 2021). Moreover, centralised management structures are susceptible to cybersecurity threats because a central control system is vulnerable to hacking, which can compromise the entire city (Razmjoo et al. 2021). Moreso, while Noori et al. (2020) advocate for centralized governance as efficient, Majeed et al. (2021) and Khan et al. (2020) caution against risks of opacity and reduced responsiveness.

3.4.2 Decentralised Governance Models

The decentralised model is the opposite of the centralised model, which involves delegating power to various organisational members. As pointed out by Ismagilova et al. (2022), decentralised governance entails a form of organisation of authority that promotes mutual responsibility among central, regional, and local institutions by the principle of subsidiarity. This approach enhances the quality, efficiency, and capacity of governance, thereby encouraging innovation and integrating smart city efforts with community priorities (Bokhari and Myeong 2023). It enables community empowerment by developing specific neighbourhood policies that result in constructive approaches, such as community-based initiatives (Lai et al. 2020). Decentralisation ensures innovation and promotes change by saving time spent negotiating with central authorities. Although decentralised governance models present a possible solution to smart cities' challenges, the problems of coordination, resource scarcity, infeasible economies of scale, community involvement, innovation, and adaptability cannot be denied (Khan et al. 2020).

3.4.3 Hybrid Governance Models

The Hybrid Governance Model is a governance model that combines elements of both authoritarianism and democracy, allowing for the potential coexistence of political repression and regular elections (Majeed et al. 2021). Various hybrid governance forms characterise most smart city governance, as it utilises both centralised and decentralised governance mechanisms (Bilal et al. 2021). For instance, central control establishes general principles while local requirements are applied, balancing efficiency and responsiveness to coordinate efforts in sustainability, economic development, and technological innovation. Although hybrid models enable better resource management, collaboration, and innovation, they can only be implemented with careful design, coordination, communication, power balance, data integration, interoperability, and robust security systems (Repette et al. 2021). For easy flow and identifaction, Table 3.1 provides summary for the governanace models.

3.5 The Role of Stakeholders in Smart Cities Governance (SCG)

Stakeholders in SCG are individuals, groups, agencies, parties, or organisations involved in the context of SCG, regardless of their level of involvement. Blasi et al. (2022) noted that they play a crucial role in the effectiveness of SCG. This section looks at the responsibilities of some of these stakeholders.

3.5.1 Government Agencies and Departments

One of the key players in the governance of smart cities is the government agencies and departments whose central role is to provide the framework, resources, and overall management requirements for these projects. He et al. (2022) explain that they are tasked with creating policies, enforcing standards, and regulating activities in various sectors, including transportation, energy, safety, and communication. Smart cities cannot operate without government intervention in their development and deployment, given the prevailing focus on public interest and social gain. Additionally, they are tasked with the mandate of knowledge development, capitalisation, and strategic management. Public stakeholders

Table 3.1 Comparisons of various smart city governance models

Governance model	Description	Advantages	Disadvantages
Centralised	Single authority controls data, decisions, and implementation	Rapid resource allocation, uniform policy enforcement, efficient project execution	Lack of transparency, ignores local needs, data overload risks, cybersecurity vulnerabilities
Decentralised	Power distributed across multiple levels (central, regional, local)	Encourages innovation, community empowerment, localized solutions, increased responsiveness	Coordination challenges, resource limitations, inefficiencies at scale
Hybrid	Mix of centralised efficiency and decentralised responsiveness	Balances control and flexibility, enables collaboration, efficient resource use	Requires complex coordination, robust systems, and careful design to avoid dysfunction

like being involved in policy formulation, regulation, and compliance monitoring (Bilal et al. 2021). Local government departments are responsible for implementing smart city projects within a specific locale after central government agencies define a strategic plan that guides the process, providing funding and motivation for innovation (Zhou et al. 2021). In turn, government agencies and departments are responsible for ensuring data security and privacy, as well as maintaining the integrity of their data systems (Nesti 2020).

3.5.2 Private Sector Entities

Private Sector Entities are one of the stakeholders in SCG, which primarily comprise entities operating in the private sector, such as multinationals, companies, start-ups, businesses, and vendors. According to Jiang (2021), these stakeholders play a critical role in SCG as they pursue economic and commercial success and market share and fulfil their corporate citizenship responsibilities. For example, ICT businesses are responsible for providing effective hardware and software products, information systems, solutions, and services to the population and its stakeholders (Zhou et al. 2021). Private sector entities also help close the financing gap between public funding and capital for large smart city projects. For instance, financial institutions can provide innovative financial solutions, such as PPPs and green bonds, that enable cities to plan ambitious projects within their budgets without burdening the public purse (Qin and Qi 2021).

3.5.3 Civil Society Organizations (CSOs)

Nastjuk et al. (2022) acknowledge the importance of CSOs in smart city governance and development, as they serve as community agents who promote and advocate for inclusive policy initiatives and public participation. These organisations include Non-Governmental Organisations, community (NGO's) groups, advocacy organisations, and academic institutions that are committed to ensuring the smart city is inclusive, sustainable, and citizen-centred. De Guimarães et al. (2020) added that CSOs act as advocates for vulnerable people in smart cities to see that smart city technologies and programs do not widen the inequality gap. In addition, they promote diversity in infrastructure and technology, and those focusing on social inequality encourage affordable housing and access to digital products and services (Oliveira et al. 2020). Finally, CSOs play a crucial role in the administration of smart cities, as they oversee the project execution process and ensure that government activities align with stated goals and ethical standards (Nastjuk et al. 2022).

3.5.4 Citizens and Community Groups

As rightly noted by Grossi et al. (2020), citizens and communities are critical stakeholders in SCG, as they ensure that these smart projects focus on real needs and are effective, transparent, and inclusive. As argued by Kirimtat et al. (2020), citizen involvement helps diffuse trust and legitimacy for smart city projects, creating a sense of ownership and community support. Community groups, such as neighbourhood associations, local advocacy groups, and volunteer organisations, mediate between individual citizens and city authorities by mobilising residents, organising meetings and conversations on various topics, and fostering people's interest in governmental processes (Oliveira et al. 2020). According to Masik et al. (2021), organising and participating in public forums helps reach a wider audience and ensures that their views are incorporated into the smart city projects being developed. Finally, community groups also exist to promote and protect the interests of their members, ensuring that smart cities do not end up empowering some groups over others (Lim et al. 2023).

3.5.5 Collaborative Partnerships and Networks

Kuzior et al. (2023) emphasise the importance of partnerships and networks in smart city governance. These kinds of partnerships involve various stakeholders, including public sector regulatory frameworks and policies, as well as the development of private sector knowledge and financial power. For instance, a project to enhance air quality may involve city planners, environmental NGOs, the creators of the pollution measurement tool, and a university that conducts pollution research (Xiao and Xie 2021).

3.6 Challenges and Opportunities in Smart Cities Governance

Even though smart governance is a critical aspect in the effective implementation of innovative city policies, it is fraught with challenges and opportunities, including:

3.6.1 Data Privacy and Security Concerns

Today, the concept of smart cities is not only dependent on precise and wide-scale data but also uses it to provide better urban amenities, increase efficiency, and maximise quality of life. Nonetheless, this data-driven model raises significant concerns regarding privacy and cybersecurity. The sheer volume of data being constantly produced by sensors, surveillance cameras, and connected devices is exposing a vulnerable digital infrastructure (Choenni et al. 2022). A relevant case is that of Amsterdam, where a Privacy by Design regime has been implemented to inform data collection policies, including anonymisation processes, open data transparency instruments, and the citizen consent program (Ismagilova et al. 2022). These efforts aim to mitigate risks and foster confidence in government agencies.

Conversely, the inability to protect this type of data might lead to major breaches. The 2017 ransomware attack on Atlanta's municipal systems crippled various city services and exposed the vulnerabilities of the local digital defences, underscoring the necessity of cybersecurity preparedness within the context of urban governance. Other cities, such as Barcelona, have implemented end-to-end encryption and multi-layer authentication on critical infrastructure platforms in response to the same threats (Oliveira et al. 2020). Additionally, legislative tools like the General Data Protection Regulation (GDPR), accepted by the EU, have established worldwide standards of data protection, contributing to the importance of accountability, user rights, and legal redress practices. As such, to earn the trust of all people, cities can implement not just technical security measures but also governance modes committed to transparency, guaranteeing accountability, and offering mechanisms of participation in supervision (Nastjuk et al. 2022). In their absence, the smart city initiatives can run the risk of alienating the citizens whom they are supposed to empower.

3.6.2 Interoperability and Standardisation Issues

The core concepts behind successful smart city execution are interoperability and standardisation, as they enable different systems, technologies, and data formats to be integrated cohesively (Oliveira et al. 2020). The absence of these leads to disjointed infrastructure, data silos, and inefficiencies in cities caused by non-communicative platforms. Isolated systems hinder real-time decision-making and may negatively impact the delivery of urban services, as illustrated by early smart grid projects in the United States, where interconnections between proprietary systems of different suppliers were expensive and inefficient (Kirimtat et al. 2020).

The FIWARE framework is a notable example of a solution to this challenge developed within the European Union. FIWARE is an open-source set of APIs and data models that support interoperability among industry verticals, including transport, energy, and public services. It has already been successfully implemented in the cities of Santander, Spain and Porto, Portugal, to integrate sensor data and service delivery platforms (Noori et al. 2020). Moreover, including legacy infrastructure, which may require significant amounts of time and cost, is mitigated using modular architecture and standardised information protocols (Grossi et al. 2020). Thus, digital cities that seek scalable and future-proof digitalisation should pursue open standards, be willing to enable cross-sector collaboration and invest in interoperability by design. These provide not only cost efficiency but also enhance the coordination of stakeholders in intelligent governance (Estevez et al. 2021).

3.6.3 Regulatory Compliance and Legal Frameworks

A smart city government is incomplete without regulatory compliance, which ensures that technologies are developed within the legal, ethical, and social realms of expectation (Razmjoo et al. 2022). Nevertheless, the complexity of regulatory environments often can impede the development of smart cities due to mismatched approaches to data governance, cybersecurity, and accountability to its citizens. These discrepancies hinder stakeholder understanding and make inter-jurisdictional cooperation more challenging (Lytras and Şerban 2020).

For example, the European Union has the General Data Protection Regulation (GDPR), which provides a comprehensive legal framework within the EU, regulating data use, privacy, and consent across all member states. Helsinki and Vienna are cities that have introduced smart city platforms compliant with the GDPR, featuring privacy-by-design aspects and allowing citizens to control digital services (Abdel-Razek 2021). These models demonstrate how regulation can foster innovation and accountability by providing clarity. However, the correlation between regulation and technological potential is an intricate one. Excessive regulations can suppress innovation, whereas insufficient rules can lead to unregulated monitoring, algorithm bias, and a loss of trust among people (Oliveira et al. 2020). Hence, visionary cities must strike a balance by creating flexible legal frameworks that adapt to emerging technologies. The government, academics, and other innovators should collaborate in the policymaking process to create legal standards that not only preserve rights but also promote transformative urban solutions.

3.6.4 Funding and Resource Allocation

Financial viability is a key aspect of smart city governance, where funds are allocated to maintain a city and enhance its overall quality of life. However, potential barriers include increased treatment costs, competition for funding, and unequal access to funding (Soeiro 2021). This sometimes means that limited public funding results in necessary projects being funded externally, while increased competition from other sectors, such as healthcare, education, and public safety, also affects the funding of infrastructure projects (Grossi et al. 2020). Such an imbalanced allocation of funds may exacerbate existing disparities between affluent and impoverished urban communities. These problems require projects that address underserved populations and build inclusive communities to achieve higher grant award rates (Humayun et al. 2020).

3.6.5 Equity and Social Inclusion

Smart city governance is governance that targets combating inequality and ensuring equal opportunities for all citizens, but difficulties include digital exclusion that affects poor households, racial minorities, older adults, and disabled people and ethical concerns like data bias and discrimination (Ooms et al. 2020). For instance, data on which smart city technologies are based is often incomplete or misinterpreted, perpetuating existing injustices and hindering the development of equity and social inclusion (Polese et al. 2021). Furthermore, Masik et al. (2021) highlighted the issue of gentrification and displacement, which may arise if smart city development is not conducted in a sustainable manner. In this regard, Smart cities should target vulnerable populations to guarantee decentralised and inclusive development. For instance, inclusive outreach and community events can help engage residents and encourage political participation while supporting disadvantaged groups (Abutabenjeh et al. 2022).

3.6.6 Public Participation and Engagement

Engagement of the people is one of the cornerstones of smart city governance, ensuring that technology solutions are not only fitting but also align with the needs and values of the people. Nonetheless, inclusive participation is an essential issue. Digital illiteracy, socio-economic inequality, and language complexity frequently isolate marginalised communities on civic tech platforms, severely reducing the democratic potential of smart city projects (Kutty et al. 2020). As an example, older populations, lower-income groups, and minority language communities often have trouble using or comprehending online services, giving them little opportunity to work in policy feedback cycles (Gracias et al. 2023).

To mitigate such problems, several cities, including Barcelona, have introduced Decidim, an open-source collaborative democracy software that enables citizens to propose, debate, and co-create city policies through citizen-friendly, multilingual interfaces (Nastjuk et al. 2022). The voting app in Seoul is also a public decision-making application based on mobile media aimed at enhancing accessibility, providing notifications, and facilitating easier voting on municipal affairs (Dana et al. 2022). Successful involvement also involves creating care plans to foster trust and reduce cynicism. Masik et al. (2021) demonstrated that responsiveness and accountability can be effectively illustrated through clear communication, feedback processes, and the enforcement of citizen participation. Additionally, integrating participatory design workshops and offline connections with underserved communities is a way to bridge the digital divide. Ultimately, participatory systems will enhance urban resilience and contribute to making smart city development more equitable and citizen-focused.

3.6.7 AI and Smart City Governance

The real-time optimisation of services, predictive modelling, and data-driven decision-making are rapidly transforming the current state of smart cities through the adoption of AI. AI technologies allow an effective system response and recovery by supporting the allocation of resources, management of mobility, and the predictive maintenance of systems when combined with urban IoT structures (Bilal et al. 2021; Zhou et al. 2021). Artificial intelligence-based platforms enable the analysis of high volumes of real-time data flows, allowing city managers to gain an understanding of energy consumption, waste routes, and citizen activity (Gracias et al. 2023). Additionally, AI helps digitalise governance regimes, automating bureaucratic procedures and facilitating citizen participation boards (Qin and Qi 2021). Nevertheless, these advantages are accompanied by notable governance risks, such as algorithmic opacity, data bias, and ethical issues related to surveillance and accountability (Nastjuk et al. 2022; De Guimarães et al. 2020). The strong dependence on AI systems without straightforward and understandable models poses a risk of harming public trust and exacerbating pre-existing disparities in digital access and service delivery (Kutty et al. 2020). In this way, ethical protection and control systems are to be deployed in smart city governance structures to promote equitable and inclusive adoption of AI (Majeed et al. 2021). Indeed, places like Amsterdam and Seoul have also established participatory models of AI governance as an alternative (Ooms et al. 2020). Finally, sustainability, inclusivity, and the ability to connect with the values of human life and its overall improvement in the context of urban development must also drive the incorporation of AI into urban governance (Dana et al. 2022; Masik et al. 2021).

3.7 Best Practices and Lessons Learned

In this section, you will find useful information about the best practices in the governance of smart cities. Singapore's remarkable success story and its adoption of a people-centric strategy in smart city governance, as seen in Barcelona, will be examined further. Additionally, the sustainable planning and design of the city of Copenhagen will be discussed in the next subsection. Moreover, you can find additional information on data infrastructure and privacy to advance policymaking with data analytics, which was implemented by the city of Amsterdam.

3.7.1 Success Stories in Smart Cities Governance

The governance of smart cities has been a success story in many countries as radical reforms have improved city life. For instance, Singapore is cited as a success story in smart city governance. It has adopted smart governance and developed a comprehensive approach to urban planning and governance, utilising technology to address various urban issues. Dana et al. (2022) provided an example of a smart city initiative in Singapore, where the Smart Nation Sensor Platform integrates data from various sectors to support decision-making, resource allocation, and service delivery. Once again, Barcelona has adopted a people-centric approach in its smart city governance, consistently acting in favour of its citizens' needs and wishes. The city has implemented several measures to enhance the quality of life and promote citizen participation. For example, the Barcelona Participa platform enables citizens to collaborate with the city's authorities to develop solutions that enhance the transparency and accountability of smart city solutions, contributing to social inclusion and facilitating social cohesion (De Guimarães et al. 2020). Copenhagen has also developed into a sustainable city focusing on environmental sustainability, the attractiveness of living, and resilience (Ullah et al. 2021). One example is Copenhagen, which focuses on sustainable planning and design of the city, active transportation, and energy distribution; this has reduced carbon emissions and improved public health and resilience (Ooms et al. 2020). In addition, Amsterdam has adopted evidence-based management to enhance urban operations, services, and the quality of life for its residents. Amsterdam invests in data infrastructure and privacy to advance policymaking through data analytics, increasing efficiency and reducing costs while providing customised services (Maccani et al. 2020).

3.7.2 Strategies for Overcoming Governance Challenges

Drawing on the success stories of smart city governance in various cities around the world, it is essential to highlight strategies for overcoming governance challenges in some cities. Figure 3.4 shows the Strategies that stem from the successes of their applicability in some parts of smart cities.

Figure 3.4 demonstrates that Strategies for Overcoming Governance Challenges include developing cross-sectoral collaboration. Nesti (2020) asserted that building relationships that span sectors, from public to private entities, will help cities identify a variety of competencies, resources, and perspectives necessary to address complex issues. Cooperation enables the sharing of resources, expertise, and problem-solving skills, leading to practical improvements and innovation.

Transparency and accountability, key pillars of governance, are further bolstered by cross-sectoral collaboration. It recommends that cities increase their transparency in decision-making, including involving stakeholders and establishing institutional mechanisms that support citizen engagement (He et al. 2022). Community

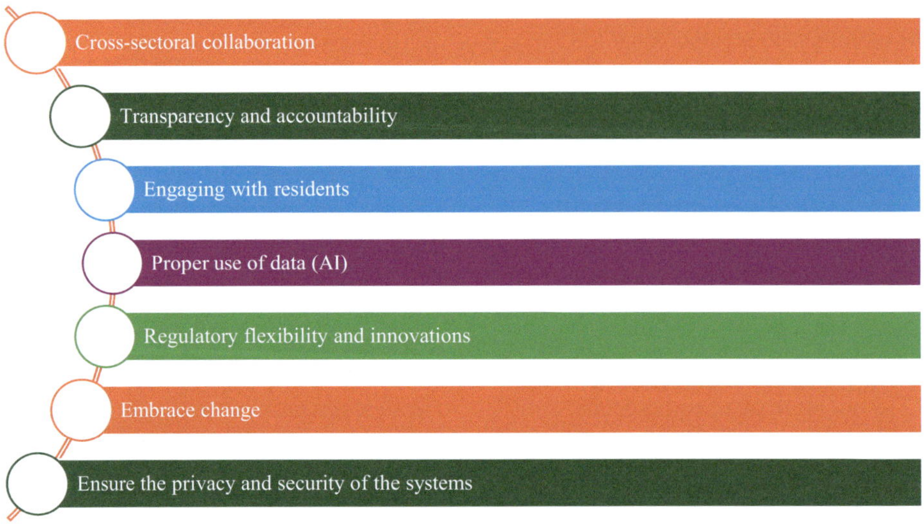

Fig. 3.4 Strategies for overcoming governance challenges. (Source: Author's compilation)

empowerment is also important to governance because it brings residents into the decision-making and designing process (Nastjuk et al. 2022). Cities can utilise local knowledge and imagination in urban planning in ways that are more reflective of residents' needs and dreams (Qin and Qi 2021). Lai et al. (2020) recommend the application of data analytics, AI, and machine learning to support planning, transport, services, and other facets of a city. Furthermore, Blasi et al. (2022) recognised regulatory freedom and innovations as crucial for adopting new technologies and reporting new developments for testing and trials. Repette et al. (2021) demonstrated the necessity for governments and all members of the smart city network to prioritise innovation and learning to foster the emergence of creativity, agility, and flexibility, enabling cities to respond to change and capitalise on opportunities in smart city governance.

3.7.3 Recommendations for Future Governance Initiatives

Sustainability, technological developments, rapid urbanisation, and smart urban management and governance have made smart city governance vulnerable. For instance, the COVID-19 pandemic widened the digital divide and necessitated the establishment of infrastructure and the affordability of the internet, as well as initiatives for teaching digital competencies (Zhou et al. 2021). Other areas of need include climate action and resilience, where the cities deploy smart transportation, renewable energy, and green building measures. Technologies like AI, IoT, and big data analytics are revolutionising people's lives—as avenues for cost reduction and efficient resource utilisation, as well as risk assessment and addressing issues of privacy, cybersecurity, and algorithmic bias (Bilal et al. 2021). On this note, future governance plans should shift from the current focus on the economic sphere to a broader consideration of the social, economic, and ecological issues. This includes engaging various government agencies and collaborating with other stakeholders to address complex urban environmental issues (Khan et al. 2020).

It is also important that equity and inclusion be prioritised in future nationally crafted governance agendas that will embrace the use of smart city technologies for all citizens, regardless of their economic status. This means that efforts must eliminate technology access and support that is biased (Bokhari and Myeong 2023). More important is the high volume of data cities collect and use to make decisions about privacy concerns, security, information bias, and discrimination risks with data governance interventions. As explained by Lai et al. (2020), this should be achieved by creating clear policies for data collection and storage, as well as for sharing and utilisation, to ensure transparency and accountability.

Subsequent governance transitions should encourage civic engagement in good faith to ensure that the people are involved in the change process. This implies the need to involve residents, gather feedback, and empower them to create solutions that align with their preferences through collaboration (Razmjoo et al. 2021). All future governance initiatives must foster resilience, promote conditions for innovation, enhance organisational capabilities for understanding and utilising technology and digital data, and encourage the application of environmentally friendly technologies, as well as risk-taking and digital literacy (Majeed et al. 2021).

3.8 Conclusion

This chapter emphasises the importance of governance in driving the growth and transformation of smart cities, with a focus on inclusive, transparent, and adaptive governance systems. It highlighted how different stakeholders, including government agencies, the private sector, citizen-oriented organisations, and individuals, collaboratively determine urban outcomes through common policymaking, innovation, and regulatory efforts. The most fundamental governance issues, including data privacy, interoperability, regulatory alignment, and fair community input, were discussed, and strategies to address them are provided for the real world. A set of successful case studies, such as those in Singapore, Barcelona, and Amsterdam, has confirmed that smart governance can be implemented effectively to achieve sustainable and resilient urban systems. With a focus on the future, technological predictions surrounding smart cities involve AI, data governance, public-private partnerships (PPPs), and flexible regulatory models. However, these should be informed by ethical practices, social inclusions, and civil trust. Finally, future governance in smart cities has been proposed to be guided by a roadmap and maturity model, addressing all the involved governance issues on foundational planning, advanced technology integration, stakeholder engagement, and key tensions. The top technological trends to watch are AI, data governance, PPPs, and flexible regulatory frameworks, and have to be steered by ethical practices, social inclusion, and civic trust. Collecetively, cities must combine innovation, transparency, legal protection and active engagement in other to prosper.

References

Abdel-Razek SA (2021) Governance and SDGs in smart cities context. In: Smart cities and the un SDGs. Elsevier, pp 61–70

Abutabenjeh S, Nukpezah JA, Azhar A (2022) Do smart cities technologies contribute to local economic development? Econ Dev Q 36(1):3–16

Bilal M, Usmani RSA, Tayyab M, Mahmoud AA, Abdalla RM, Marjani M, Pillai TR, Targio Hashem IA (2021) Smart cities data: framework, applications, and challenges. In: Handbook of smart cities. Springer, pp 113–141

Blasi S, Ganzaroli A, De Noni I (2022) Smartening sustainable development in cities: strengthening the theoretical linkage between smart cities and SDGs. Sustain Cities Soc 80:103793

Bokhari SAA, Myeong S (2023) The influence of artificial intelligence on e-Governance and cybersecurity in smart cities: a stakeholder's perspective. IEEE Access

Choenni S, Bargh MS, Busker T, Netten N (2022) Data governance in smart cities: challenges and solution directions. J Smart Cities Society 1(1):31–51

Dana LP, Salamzadeh A, Hadizadeh M, Heydari G, Shamsoddin S (2022) Urban entrepreneurship and sustainable businesses in smart cities: exploring the role of digital technologies. Sustain Technol Entrep 1(2):100016

De Guimarães JCF, Severo EA, Júnior LAF, Da Costa WPLB, Salmoria FT (2020) Governance and quality of life in smart cities: towards sustainable development goals. J Clean Prod 253:119926

Estevez E, Pardo TA, Scholl HJ (eds) (2021) Smart cities and smart governance: towards the 22nd century sustainable city, vol 37. Springer Nature

Gracias JS, Parnell GS, Specking E, Pohl EA, Buchanan R (2023) Smart cities—a structured literature review. Smart Cities 6(4):1719–1743

Grossi G, Meijer A, Sargiacomo M (2020) A public management perspective on smart cities: 'Urban auditing'for management, governance and accountability. Public Manag Rev 22(5):633–647

He W, Li W, Deng P (2022) Legal governance in the smart cities of China: functions, problems, and solutions. Sustainability 14(15):9738

Humayun M, Jhanjhi NZ, Alamri MZ, Khan A (2020) Smart cities and digital governance. In: Employing recent technologies for improved digital governance. IGI Global, pp 87–106

Ismagilova E, Hughes L, Rana NP, Dwivedi YK (2022) Security, privacy and risks within smart cities: literature review and development of a smart city interaction framework. Inf Syst Front:1–22

Jiang H (2021) Smart urban governance in the 'smart' era: why is it urgently needed? Cities 111:103004

Khan HH, Malik MN, Zafar R, Goni FA, Chofreh AG, Klemeš JJ, Alotaibi Y (2020) Challenges for sustainable smart city development: a conceptual framework. Sustain Dev 28(5):1507–1518

Kirimtat A, Krejcar O, Kertesz A, Tasgetiren MF (2020) Future trends and current state of smart city concepts: a survey. IEEE Access 8:86448–86467

Kutty AA, Abdella GM, Kucukvar M, Onat NC, Bulu M (2020) A system thinking approach for harmonizing smart and sustainable city initiatives with United Nations sustainable development goals. Sustain Dev 28(5):1347–1365

Kuzior A, Pakhnenko O, Tiutiunyk I, Lyeonov S (2023) E-governance in smart cities: global trends and key enablers. Smart Cities 6(4):1663–1689

Lai CS, Jia Y, Dong Z, Wang D, Tao Y, Lai QH, Wong RT, Zobaa AF, Wu R, Lai LL (2020) A review of technical standards for smart cities. Clean Technol 2(3):290–310

Lim Y, Edelenbos J, Gianoli A (2023) Dynamics in the governance of smart cities: insights from South Korean smart cities. Int J Urban Sci 27(sup1):183–205

Lytras MD, Şerban AC (2020) E-government insights to smart cities research: European Union (EU) study and the role of regulations. IEEE Access 8:65313–65326

Maccani G, Connolly N, McLoughlin S, Puvvala A, Karimikia H, Donnellan B (2020) An emerging typology of IT governance structural mechanisms in smart cities. Gov Inf Q 37(4):101499

Majeed U, Khan LU, Yaqoob I, Kazmi SA, Salah K, Hong CS (2021) Blockchain for IoT-based smart cities: recent advances, requirements, and future challenges. J Netw Comput Appl 181:103007

Masik G, Sagan I, Scott JW (2021) Smart City strategies and new urban development policies in the Polish context. Cities 108:102970

Nastjuk I, Trang S, Papageorgiou EI (2022) Smart cities and smart governance models for future cities: current research and future directions. Electron Mark 32(4):1917–1924

Nesti G (2020) Defining and assessing the transformational nature of smart city governance: insights from four European cases. Int Rev Adm Sci 86(1):20–37

Noori N, Hoppe T, de Jong M (2020) Classifying pathways for smart city development: comparing design, governance and implementation in Amsterdam, Barcelona, Dubai, and Abu Dhabi. Sustainability 12(10):4030

Oliveira TA, Oliver M, Ramalhinho H (2020) Challenges for connecting citizens and smart cities: ICT, e-governance and blockchain. Sustainability 12(7):2926

Ooms W, Caniëls MC, Roijakkers N, Cobben D (2020) Ecosystems for smart cities: tracing the evolution of governance structures in a Dutch smart city initiative. Int Entrep Manag J 16:1225–1258

Polese F, Troisi O, Grimaldi M, Loia F (2021) Reinterpreting governance in smart cities: an ecosystem-based view. In: Smart Cities and the UN SDGs. Elsevier, pp 71–89

Qin B, Qi S (2021) Digital transformation of urban governance in China: the emergence and evolution of smart cities. Digital LJ 2:29

Razmjoo A, Østergaard PA, Denai M, Nezhad MM, Mirjalili S (2021) Effective policies to overcome barriers in the development of smart cities. Energy Res Soc Sci 79:102175

Razmjoo A, Gandomi AH, Pazhoohesh M, Mirjalili S, Rezaei M (2022) The key role of clean energy and technology in smart cities development. Energ Strat Rev 44:100943

Repette P, Sabatini-Marques J, Yigitcanlar T, Sell D, Costa E (2021) The evolution of city-as-a-platform: smart urban development governance with collective knowledge-based platform urbanism. Land 10(1):33

Soeiro D (2021) Smart cities and innovative governance systems: a reflection on urban living labs and action research. Fennia Int J Geograp 199(1):104–112

Tan SY, Taeihagh A (2020) Smart city governance in developing countries: a systematic literature review. Sustainability 12(3):899

Ullah F, Qayyum S, Thaheem MJ, Al-Turjman F, Sepasgozar SM (2021) Risk management in sustainable smart cities governance: a TOE framework. Technol Forecast Soc Chang 167:120743

Xiao X, Xie C (2021) Rational planning and urban governance based on smart cities and big data. Env Technol Innov 21:101381

Zhou Q, Zhu M, Qiao Y, Zhang X, Chen J (2021) Achieving resilience through smart cities? Evidence from China. Habitat Int 111:102348

Dynamics of Smart Cities in Developed and Developing Economies

4

Abstract

This chapter investigated the contrasting trajectories of smart city development in developed and developing economies, situating these within the broader context of rapid global urbanisation. It highlighted that while smart city initiatives often prioritise sustainability, governance, and quality of life, their success hinges on economic capacity, infrastructure, and coherent policy frameworks. Case studies from cities such as Singapore, London, and Amsterdam also exemplified advanced data-driven models but revealed concerns over digital divides and privacy issues. Conversely, in developing regions such as Africa, South Asia, and Latin America, the chapter showed that implementation is hampered by inadequate infrastructure, reliance on external funding, and political instability. Innovative, low-cost solutions exist but often fall short of addressing deeper structural governance issues.

Moreover, the chapter critiqued the uncritical transfer of smart city models from the Global North to the Global South, arguing for locally tailored approaches. It also examined emerging technologies, including AI, IoT, and blockchain, as well as their ethical implications. Consequently, the chapter concludes by advocating for inclusive, adaptive, and ethically governed smart city frameworks that prioritise human development over purely technological advancement.

4.1 Introduction

The twenty-first century has transformed urban areas from physical spaces into interconnected digital systems that operate through data platforms, utilising automation and connectivity. As such, the Urban population surged from 39% in 1980 to 54% in 2015 and is expected to reach 66% by 2050, requiring smart city approaches to address congestion, pollution, governance, and economic gaps (Urbanet 2025). Specifically, a smart city is an urban setting where technological and data solutions enhance city operations and sustainability while improving resident living standards (Moura and de Abreu e Silva 2021). Expanding on this, Iryna et al. (2022) defined it as maintaining traditional networks and services alongside information technology, which creates more adaptable systems that utilise digital technology to enhance city operations for residents. Addas (2023) presents a six-section framework for smart cities, encompassing smart governance, smart economy, smart infrastructure, smart mobility, smart environment, and smart living—a concept first introduced in the 1990s.

V. Yorucu et al., *Modern Approach of Resilient and Sustainable Smart Cities*, Sustainable Development Goals Series, https://doi.org/10.1007/978-3-032-05715-0_4

Currently, the importance of Smart Cities in global urban development is essential, particularly in the context of rapid urbanisation, climate change, and economic instability. For instance, a worldwide population prediction for 2050 indicates that 66% of humanity will reside in cities, alongside growing demands for efficient urban services and infrastructure (Statista 2025). As such, smart technologies will enable cities to improve energy efficiency, reduce congestion, and enhance public safety, thereby supporting sustainability and urban resilience (De Falco 2019). Smart cities have demonstrated significant financial importance as they generate innovation and investment opportunities. For instance, the worldwide Smart Cities market is expected to experience significant revenue growth until 2025, reaching an estimated US$79.94 billion, driven by strong government and private stakeholder interest in smart city initiatives, which is projected to total $189.5 billion by 2025 (Statista 2025). In addition, another important aspect of smart cities in global urban development is that they promote social inclusion through their support of digital competency and their involvement of citizens in governance processes (Kummitha and Crutzen 2017). Ultimately, the concept encompasses vital elements of sustainability, emphasising the importance of efficient resource management, energy conservation, and active community engagement (Khan et al. 2020). However, the evolution of smart cities has unique pathways of urban development between developed and developing economies, along with their respective challenges and the pace and focus of urban progress.

According to Angelidou et al. (2018), smart cities in developed countries, such as the United States, Japan, the European Union, and Switzerland, are led by the capital of developed countries, characterised by advanced digital infrastructure, strong governance frameworks, and robust investment capabilities. Additionally, the connectivity in these cities is high-speed; they have intelligent urban governance and clean power, which makes them much more competitive and livable than other cities. On the other hand, smart city development in the developing economies of Africa, South Asia, and Latin America faces its own set of barriers, including inadequate infrastructure, poor governance, and limited financial resources (Hatuka et al. 2018). Additionally, they must navigate digital divides, informal settlements, and political instability; therefore, they require international funding and foreign investment to make any progress (Moura and de Abreu e Silva 2021).

Based on the foregoing, this chapter considers the dynamics of smart cities in developed and developing economies. This study examines the evolution of smart cities, their core components, success factors, and implementation in developed and developing economies, including North America, Europe, Asia, and the use of emerging technologies. Ultimately, the chapter seeks to advance the discourse on urban transformation by providing strategic insights for policymakers, urban planners, and technology stakeholders.

4.2 Conceptual Framework of Smart Cities

Cities have always been one of humankind's great inventions, technological test beds and one of the greatest labs for innovation and progress from which to create the modern world. The concept of smart cities originated in the early twentieth century. However, it came to be referred to by the term in the late twentieth and early twenty-first centuries. For instance, the theory of smart cities can be traced through different phases, during which distinct concepts, technological innovations, and urbanisation trends were presented (De Falco et al. 2019). For instance, cities throughout history have been the engines that drive economic growth and cultural development; however, there is an increasing trend of congestion, pollution, and a poor resource footprint in cities. In this way, it can be concluded that it involves a transformation from traditional reactor urban planning to proactive, technology-based urban planning, aiming to enhance efficiency, sustainability, and the well-being of citizens (Attaran et al. 2022).

Concurrently, before the birth of digital technology, urban centres were planned according to industrialisation and the 'basis' of these economic expansions. The planning of cities in the early twentieth century was based on a network of transportation facilities (roads and railways) with a low degree of technology integration (Ristvej et al. 2020). According to Sharifi (2019), there was very little governance in the form of bureaucracy, with manual systems still being utilised, resulting in no real-time public responsiveness to its services.

During the 1990s, the increased use of ICT led to a new form of urbanisation known as digital urbanisation, which facilitated the better management of urban services (Ismagilova et al. 2019). For instance, cities were able to shift from traditional infrastructure to dynamic, communicative ones by adding smart grids, digital governance platforms, and automated transportation systems. The concept has been advanced by the Internet of Things (IoT), Artificial Intelligence (AI), and big data analytics, which enable the tracking of urban functions in every aspect of real-time and optimise them in line with real needs (Komninos et al. 2019). Recently, this has been demonstrated in several cities in both developed and developing countries. In particular, cities like Singapore, Barcelona, and Tokyo in developed markets serve as examples of how fundamental smart city aspects can be applied in policy, mobility, and infrastructure. Singapore has deployed predictive analytics in traffic and government services, Barcelona presents IoT-enhanced waste management, and Tokyo has deployed AI-driven elderly care (Aurigi and Odendaal 2022). Innovative solutions are emerging in developing economies, including national-scale solutions and frugal approaches, as seen in Rwanda and Kenya. Digital transformation is evident in Rwanda's national e-governance strategy in Kigali, as well as the city's smart transport initiatives, despite its limited infrastructure. M-Pesa in Kenya is an illustration of FinTech facilitating financial inclusion in environments characterised by low traditional banking penetration (Bauer et al. 2021). These examples further demonstrate how smart city core elements, as exemplified in Fig. 4.1, are adaptable to diverse development settings despite the highly differentiated problems and resource capacities.

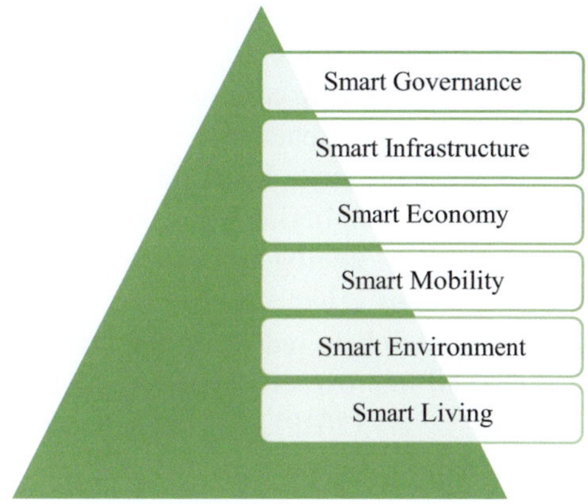

Fig. 4.1 Components of a smart city. (Source: Author's compilation)

As a key component of a smart city, smart governance leverages digital technologies to enhance decision-making, service delivery, and public participation (Artmann et al. 2019). As Zheng et al. (2020) proposed, reducing bureaucratic inefficiencies and increasing transparency make governments' administrative processes more streamlined. For example, London's e-Governance Model is an instance where 99% of services, including taxation, healthcare, and business registration, can be administered online, resulting in a decrease in paperwork and administrative inefficiency (Bai et al. 2023). In addition, the smart infrastructure of a smart city plots the digital networks, smart energy grids and automated urban systems to add efficiency and connectivity. Qayyum et al. (2021) state that intelligent infrastructure is beneficial for managing various services, such as electricity, water, and waste, more efficiently and effectively by utilising sustainable resources. Songdo City in South Korea is an example of smart infrastructure, where all IoT sensors enable maximum energy consumption, energy mobilisation through an automated waste disposal process, and energy sustainability (Attaran et al. 2022).

A smart city's smart economy component further entails using digital technologies to enhance economic productivity, creativity, and financial inclusion, thereby promoting the economic participation of individuals. To illustrate, Ristvej et al. (2020) mention the contribution of FinTech to promoting entrepreneurship in developing countries, such as Kenya, by facilitating cashless transactions and simplifying business operations. In smart mobility, intelligent transportation systems, autonomous vehicles, and digital ridesharing are utilised to help reduce congestion, lower emissions and improve efficiency. According to Ismagilova et al. (2019), the management of Singapore's traffic system demonstrates the environmental and economic benefits of transportation provided to citizens within a well-integrated mobility system.

Moreover, a smart environment, as part of a smart city, is a crucial component in creating smart cities, particularly in terms of energy efficiency, waste management, and pollution regulation. This is because climate change and rapid urbanisation have intensified environmental challenges, making smart and data-driven solutions necessary (Komninos et al. 2019). For example, smart sensors are used to sense waste levels and optimise collection routes in implementing Barcelona's IoT-enabled waste management system, which reduces fuel consumption and leads to cleaner, more sustainable urban environments (Zheng et al. 2020). For instance, Smart living in smart cities also utilises digital innovation to enhance healthcare, education, and quality of life, as seen in Tokyo's AI-based elderly care services, which reduce healthcare costs and improve the well-being of older people (Bai et al. 2023).

In addition, smart city development is grounded on several theories and models, such as sustainable development theory (SDT), the technology acceptance model (TAM), and the triple helix model. For example, SDT serves as a foundation for developing a smart city, emphasising the complementary goals of economic growth, environmental sustainability, and social equity. Suárez-Eiroa et al. (2019) suggest that the theory states that cities must develop long-term resource management strategies that follow developments. However, a central obstacle to this theory remains between the goals of sustainability and the practice of urban governance. For instance, Ruggerio (2021) observes that SDT frequently precedes environmental components when urban centres are growing rapidly in poverty levels. Along these lines, Davis (1989) TAM asserts that how individuals accept and use new technology is based on two leading factors: perceived usefulness and perceived ease of use (Hou et al. 2020). Habib et al. (2023) stated that the TAM is useful in analysing citizen engagement with other digital services of e-governance, such as cashless transactions and smart transportation, in the context of smart city development. TAM aims particularly at both developed urban centres, such as Amsterdam, where the use of digital technology is high, and developing environments, such as Nairobi, where people have widely adopted mobile-first services (e.g., M-Pesa) despite the lack of developed infrastructure (Bellini et al. 2022).

Nevertheless, critics of TAM assert that the approach neglects external barriers to service (Sepasgozar et al. 2019). Calzada (2020) explain that such a model is particularly well-suited for smart city growth, as urban transformation relies heavily on multi-stakeholder collaboration, and it is through working together that urban transformation can occur. Matos et al. (2019) were critical of the theory for also overlooking grassroots participation and choosing elite institutions over the local communities.

Lastly, it is also worth noting that the differences between e-governance and digital governance exist, as the two are interconnected but have varying levels of technology use in urban governance. E-governance is often used to describe the digitalisation of public service provision, including online taxation systems, licensing, and sharing of civic records, which accelerates bureaucratic efficiency and citizen access (Zheng et al. 2020; Bai et al. 2023). Conversely, the concept of digital governance encompasses a wider institutional paradigm as it includes algorithmic decision-making, digital ethics, cybersecurity, data sovereignty, and participatory digital citizenship (Calzada 2020; Komninos et al. 2019). This form of governance extends beyond the automation of services to encompass strategic planning and the regulation of urban digital infrastructures. To illustrate, although the London e-governance system is a good example of minimised administrative interaction, real digital governance would also concern data security protocols, citizen control over the use of AI, and fair access to smart technologies. To ensure that smart solutions do not undermine transparency, inclusivity, and ethical norms, more developed smart cities should implement extensive digital governance (Artmann et al. 2019; Habib et al. 2023).

4.3 Smart City Development in Developed Economies

Developed economies are regions, typically countries, with abundant wealth and resources available to their residents or citizens (Jonek-Kowalska and Wolniak 2022). In particular, smart cities in developed economies have been identified as possessing several defining features, as illustrated in Fig. 4.2.

The findings reveal a significant difference between smart cities in developing economies and those in developed countries that are determined by their specific qualities (see Fig. 4.2). Among them, one can find such issues as inadequate infrastructure, poor digital literacy, and excessive reliance on foreign funds (De Falco 2019; Filiou et al. 2023). Other major cities, such as Kigali in Rwanda, have already become early adopters of digital governance technological platforms, where smart transportation and cashless payment systems are facilitated through government-initiated strategies (Filiou et al. 2023). Nonetheless, the sustainability of these efforts is questioned because they are dependent on external investment.

Lagos, Nigeria's economic capital, has adopted smart surveillance and e-governance systems; however, its regressive infrastructure and high urban density hinder its development. In Kenya, the capital city, Nairobi, is leveraging mobile-first solutions, such as M-Pesa and the ambitious Konza Technopolis project, to transform into a tech-driven urban centre. However, bureaucratic hurdles and financial deficits in the country hinder city development (Wu et al. 2023). Mumbai has initiated metro ventures and smart grid installations in Asia, but its challenges with slums make inclusive urbanisation a significant issue. Jakarta has AI-powered traffic systems and smart waste controls, although political instability is hindering scale. Similarly, Manila prioritises disaster resilience and IoT-driven flood monitoring, yet informal settlements and poor governance also leave it behind. Latin American cities, such as Mexico City and São Paulo, utilise AI in policing and public safety but are hindered by ongoing issues of inequality, corruption, and infrastructure problems (Waqar et al. 2023). Buenos Aires scores high in digital inclusion and access strategies for waste management and is prone to economic and political instability (Shu et al. 2023). To aid this exposition, a logical compilation of the above city case studies is presented in Table 4.1, with each city's levels clearly and explicitly discussed in the text to provide alignment and conceptual representation.

Moving forward from case studies, a lot of drivers as shown in Fig. 4.3 have been shown to

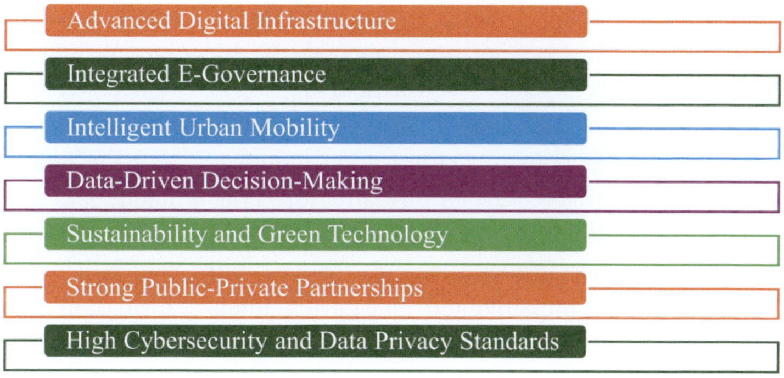

Fig. 4.2 Characteristics of smart cities in developed countries. (Source: Author's compilation)

Table 4.1 Leading smart cities

Region	City	Key smart city features	Challenges
United States	New York	• LinkNYC: Free Wi-Fi kiosks (Duygan et al. 2022) • AI-powered smart traffic management • Smart surveillance and safety	• Privacy concerns due to data collection • Limited accessibility for marginalized communities
	San Francisco	• AI-driven analytics in urban planning (Guo et al. 2022) • Blockchain integration for city management • Smart transportation initiatives (Tan and Taeihagh 2020)	• High cost of living driving out low-income residents • Socioeconomic inequality due to tech expansion
	Boston	• 311 civic engagement app for smart governance (Jonek-Kowalska and Wolniak 2022) • Predictive analytics for public services • Digital public engagement platforms	• Digital literacy • Data security concerns
Europe	Amsterdam	• Sensor-driven waste management • Autonomous energy grids • Open data and citizen participation	• Private-sector dominance in infrastructure projects
	London	• AI and IoT integration in transport • Oyster card for seamless travel (Tan and Taeihagh 2020) • Smart public safety initiatives	• Cybersecurity risks • Digital gentrification
	Barcelona	• Citizen-led smart city governance • Superblocks Project: Reduced car traffic (Guo et al. 2022) • IoT-based energy efficiency	• Potential socioeconomic exclusion in green urbanism
Asia	Singapore	• 5G smart networks • AI-driven governance • Blockchain-based e-government services (Kirimtat et al. 2020)	• Surveillance concerns and privacy issues
	Tokyo	• AI-driven elderly care and healthcare systems (Caird and Hallett 2019) • IoT-enabled urban planning • Smart transport solutions	• Rising real estate costs making smart living unaffordable
	Seoul	• Cashless public transport system • AI-powered smart traffic monitoring (Duygan et al. 2022) • Advanced digital governance	• Corporate dominance in smart city development

Source: Author's compilation

play a role in the success of smart cities in Developed Nations.

According to the findings in Fig. 4.3, high-level digital infrastructure and connectivity are drivers for smart city success. Masik et al. (2021) pointed out that smart cities in developed nations with optimal 5G networks, fibre-based connectivity, and IoT-enabled urban infrastructure contribute to seamless data collection and real-time management. For instance, Seoul, Tokyo, and Amsterdam have implemented smart grids, intelligent transportation systems, and the application of artificial intelligence in urban monitoring to enhance efficiency. Nevertheless, Digital connectivity is not inclusive, with the adoption of low-income and elderly populations being patchy, and concerns about data privacy, security vulnerabilities, and cyberattacks persist (Jonek-Kowalska and Wolniak 2021). Regarding advanced governance and policy support, smart cities in developed countries employ a progressive governance and policy framework to facilitate digital transformation. For example, cities such as Singapore, London, and Boston have advanced through

Fig. 4.3 Key Drivers of smart city success in developed nations. (Source: Author's Compilation)

AI-driven governance models, predictive analytics for public services, and blockchain-based governance (Kim 2022). However, critics argue that over-reliance on digital governance risks surveillance, and bureaucratic inefficiencies hinder the implementation of smart policies in government agencies (Duygan et al. 2022).

In addition, the success of smart cities in developed economies also hinges on strong public support, which helps sustain smart city projects through financing and development. According to Guo et al. (2022), hybrid technology companies, including Google, IBM, and Siemens, partner with governments to utilise AI-enabled urban planning, data analytics for public safety, and the development of smart infrastructure. However, corporate involvement in urban governance raises concerns about privatisation and accountability, as seen in Toronto's Sidewalk Labs smart city project, which has sparked concerns about privacy and excessive corporate control (Tan and Taeihagh 2020). Additionally, there is no evidence of AI bias or data monopoly in developed countries that invest in AI research and IoT integration for making real-time decisions (Yadav et al. 2019).

Smart cities have been successfully implemented in developed countries, as demonstrated, but some challenges have been identified that hinder their fair and sustainable growth. As an example of this problem, Hatuka et al. (2018) draw on the digital divide problem, which can lead to marginalised groups being left behind when smart city benefits are fostered. For instance, New York's LinkNYC provides free Wi-Fi; however, low-income neighbourhoods experience slower adoption rates owing to digital illiteracy and cost concerns (Angelidou et al. 2018). Moreover, cities such as London and Singapore have implemented AI-driven surveillance, which has raised ethical concerns about government overreach and mass surveillance (Khan et al. 2020). Corporate dominance in smart city projects has also been shown to pose risks of privatisation (De Falco 2019). Consequently, Smart cities must foster inclusive governance, cybersecurity resilience, ethical AI policies, and public trust to balance technological innovation with equity, ensuring transparency and citizen engagement for sustainability.

4.4 Smart City Development in Developing Economies

Developing economies are characterised by poor infrastructure, inferior growth rates, imbalanced economies, and extremely low personal incomes (Rifaid et al. 2023). By extension, smart cities in developing economies differ significantly from those in developed nations based on their unique characteristics.

As shown in Fig. 4.4, smart cities in developing economies are characterised by rapid urbanisation and population growth, which puts a strain on infrastructure. Moreover, they lack infrastructure development in basic areas such as electricity, water supply, and waste management

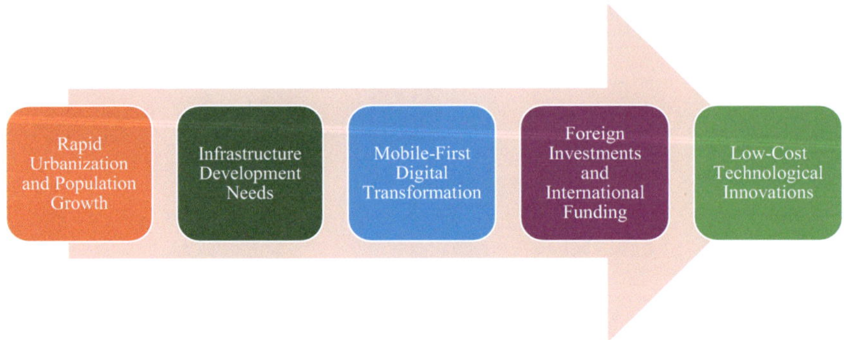

Fig. 4.4 Unique features of smart cities in developing countries. (Source: Author's compilation)

(De Falco 2019). Cities in developing economies also have low internet penetration, which drives financial inclusion through mobile-based solutions as part of a mobile-first digital transformation. Moreover, many projects in developing cities require the involvement of foreign investment and international funding, which raises concerns regarding economic dependency and geopolitical influence (Filiou et al. 2023). Lastly, low-cost technological innovations, such as solar grids and e-rickshaws, are being introduced. Additionally, Table 4.2 provides examples for easy identification of these features.

Table 4.2 provides case studies of emerging smart cities. Moving forward, studies have shown some key drivers of smart cities in developing countries, which, for clarity, are listed in Fig. 4.5.

Figure 4.5 displays the key drivers of smart cities in developing countries. As one driver of rapid urbanisation and population growth, Shu et al. (2023) documented that developing areas face exceptional urban growth, while cities such as Lagos, Mumbai, and Jakarta encounter difficulties due to the yearly migrations of numerous inhabitants. According to Chen (2023a, b), changes in population density place excessive demands on transportation systems, public services, and housing, prompting governments to deploy smart technology platforms for urban management coherence. Another driver is the need for emerging economies to struggle with implementing fundamental urban infrastructure due to their lack of reliable power systems, a clean water supply, and effective waste management regulations (Okafor et al. 2023). Another driver is that smart city construction in Africa, Asia, and Latin America relies on foreign aid, partnerships, and loans, promoting digital transformation and financial inclusion. As such, the effectiveness of these systems is limited for marginalised populations because they face vulnerabilities from cyber fraud, alongside regulatory gaps and digital illiteracy (Waqar et al. 2023).

Furthermore, developing economies face multiple hurdles in pursuing smart city initiatives. As Wu et al. (2023) pointed out, many developing countries face similar infrastructure challenges to those in developed economies, including unreliable electricity systems, insufficient transportation networks, and inferior water and sanitation systems. Bandauko and Nutifafa Arku (2023) explain that this essential failure prevents the application of smart technologies because digital solutions require reliable power systems, high-speed data networks, and optimised distribution networks. Again, developing economies face barriers to adopting smart city models, including limited access to smartphones and digital skills, and in India, Aadhaar-linked services often overlook marginalised groups (Rifaid et al. 2023).

Furthermore, the success of smart city initiatives relies on efficient governance, transparency, and effective regulatory enforcement; however, many developing countries face challenges due to political instability, inefficient bureaucracies, and widespread corruption (Hatuka et al. 2018). For instance, Jakarta's smart traffic management program has struggled due to poor executive oversight and unaccountable aspects, which have reduced the use of AI-based solutions (Yadav

Table 4.2 Case studies of emerging smart cities

Region	City	Smart city initiatives	Challenges
Africa	Kigali, Rwanda	• Smart transport systems • Cashless payments (Filiou et al. 2023) • Digital governance	• Heavy reliance on foreign investments • Government-driven control raises sustainability concerns
	Lagos, Nigeria	• Smart surveillance • Digital payment systems • E-governance	• Overpopulation and inadequate infrastructure • Eko Atlantic criticized for neglecting the broader population
	Nairobi, Kenya	• Mobile-first digital solutions (M-Pesa) (Wu et al. 2023) • Konza Technopolis ("Silicon Savannah") • AI-driven traffic systems	• Slow progress in Konza due to bureaucracy • Funding constraints limit the development
Asia	Mumbai, India	• Urban mobility projects (Mumbai Metro) (Qonita and Giyarsih 2023) • Digital governance • Smart grids for energy efficiency	• Slum integration remains a challenge • Smart projects benefit the middle class but neglect the poor
	Jakarta, Indonesia	• AI-driven traffic control • Digital waste management • Smart flood monitoring	• Political instability and corruption slow progress
	Manila, Philippines	• Disaster resilience technologies • IoT-based flood monitoring • Smart housing projects	• Informal settlements and weak governance hinder large-scale implementation
Latin America	São Paulo, Brazil	• AI-driven traffic management • Smart policing • Digital governance platforms	• High crime rates and income inequality reduce project impact
	Mexico City, Mexico	• IoT-based public safety initiatives (Waqar et al. 2023) • Open data governance • Smart mobility solutions	• Corruption and weak maintenance infrastructure • Persistent air pollution issues
	Buenos Aires, Argentina	• Digital inclusion programs • Smart waste management • AI-based urban planning (Shu et al. 2023)	• Economic instability and shifting political priorities affect long-term funding

Source: Author's compilation

Fig. 4.5 Key drivers of smart cities in developing countries. (Source: Author's compilation)

et al. 2019). Finally, rapid urban growth in cities in developing nations, as observed by Siemens (2021), leads to environmental degradation, pollution, and deforestation, surpassing the pace of green technology deployment.

4.5 Comparative Analysis: Smart Cities in Developed Vs. Developing Economies

As earlier stated, smart cities follow different developmental paths between developed and developing economic spheres. As revealed in

Research, Smart Cities in developed economies employ a data-driven approach with a structure that is focused on sustainable objectives (Matos et al. 2019). In other words, Amsterdam, Singapore, and Boston utilise integrated urban development plans that use IoT, AI and big data analytics (Calzada 2020). However, developing economies face smart city planning challenges differently, as their approach is reactive, fragmented and profoundly influenced by political and financial barriers (Sepasgozar et al. 2019). According to Habib et al. (2023), developed nations are initiating smart cities strategically with long-term approaches, whereas developing nations often suffer from ad hoc implementation due to short-term political purposes and dependence on external funding.

In developed economies, high-speed internet access and the deployment of 5G, along with established AI networks for migrating automation and governance (Hou et al. 2020), enable these technologies to achieve high digital and technological maturity. Yet, numerous countries in the developing world are confronted with low broadband capabilities, poor electricity connections, and underdeveloped technological capabilities (Ruggerio 2021). Due to this, integration with a vast range of technology is a reality for developed nations with smart cities; however, developing nations should bridge the basic digital gaps to reach similar levels of smart city advancement. Furthermore, socioeconomic circumstances and political stability are essential pillars of a smart city's success, as developed countries can enjoy steady policies and economic stability, allowing for systematic cooperation between sectors (Bellini et al. 2022). Yet, developing smart cities is hindered by significant political obstacles—instability, corruption, and socioeconomic inequalities—that prevent developing countries from making progress. For example, the smart surveillance project in São Paulo is criticised due to the governmental control that exceeds boundaries and the public's scarce transparency (Qayyum et al. 2021).

Additionally, Artmann et al. (2019) also further presented that developed nations foster smart city projects along with effective PPP between their government and businesses in private corporations. Bai et al. (2023) The private sector plays a key role in smart city development through research and development investments, as well as the deployment of technologies and infrastructure, which drives financial sustainability and innovation. However, implementing smart city projects in developing economies relies on external financial contributions, such as foreign aid, international loans, or state-controlled programs, which can lead to external dependencies and control (Zheng et al. 2020). Although developing and developed economies face different infrastructure and governance standards, they implement specific urban innovation solutions that match their local challenges. Komninos et al. (2019) demonstrated that developed nations prioritise the use of advanced technology, implementing AI automation, and sustainable urban design. For instance, Singapore and Amsterdam lead the way by adopting AI traffic surveillance technology alongside blockchain-based government systems and autonomous public transit (Qayyum et al. 2021). However, developing economies employ frugal and adaptive approaches to innovation due to financial and infrastructure limitations (Ismagilova et al. 2019). Table 4.3 comprises summarised comparisons.

4.6 Emerging Technologies and Future Prospects

The evolving state of technology defines smart cities by establishing urban districts that become more efficient through sustainable operations and better service delivery to society. The implementation of AI, blockchain technology, and IoT serves as the basic enablers for smart city systems, which offer both automated systems and security measures, as well as real-time data utilisation for informed decision-making. Springs (2024) identifies how AI systems improve traffic control procedures, predictive law enforcement methods, and environmental observation protocols, while IoT technology allows connected devices to operate cities more efficiently. Blockchain technology enables safe, decentral-

Table 4.3 Comparative analysis: smart cities in developed vs. developing economies

Aspect	Developed economies	Developing economies
Smart city planning and execution	Structured, long-term strategies integrating AI, IoT, and sustainability	Reactive, fragmented execution is often driven by political and financial constraints
Digital and technological maturity	High-speed internet, 5G, AI-driven automation, smart grids	Low digital penetration, unreliable infrastructure, digital literacy gaps
Socioeconomic and political barriers	Stable governance, strong regulatory frameworks, high economic resilience	Political instability, corruption, inequality hinder smart city progress
Public vs. private sector involvement	Strong public-private partnerships (PPPs) ensure financial sustainability	Heavy reliance on foreign aid, loans, and government-led projects
Innovation and adaptation	High-tech, sustainability-focused solutions like AI automation and blockchain	Frugal, grassroots innovations tailored to urban challenges
Urban population scale	Typically, smaller cities (e.g., Amsterdam—1.2 M, Singapore—5.9 M, London—8.9 M), enabling more manageable smart city rollouts	Mega-cities (e.g., Lagos—~20 M, Mumbai—~12.5 M) pose greater challenges due to scale, congestion, and infrastructural strain

Source: Author's compilation

ised digital transactions that increase governance transparency while enhancing data protection and financial operations in smart cities (Hussain 2024). Developing economies face significant barriers to these technological implementations due to their limited infrastructure development, cybersecurity threats, and high expense levels. Smart cities utilise smart mobility, sustainable energy, and green urbanism as emerging technologies to enhance digital transport systems and renewable energy sources, thereby addressing urban pollution, traffic congestion, and carbon emissions (Belaid et al. 2024). According to Mazzetto (2024), the adoption of electrified public transport combined with congestion pricing and AI-controlled traffic systems in Amsterdam and London resulted in notable emission reductions and decreased traffic congestion. The Global South faces challenges in finding affordable and sustainable energy alternatives, as well as local maintenance solutions, particularly in developing countries where progress has been slow. At the same time, Western smart city models continue to face implementation obstacles (Ullah et al. 2024).

In recent times, the integration of 5G networks and edge computing technologies has transformed smart cities through their ability to transmit data rapidly, offering low-latency communication and enhanced AI performance. Sharma et al. (2024) demonstrate that these technologies allow instant tracking of urban systems and control autonomous transportation, public safety networks, and industrial factory management. For instance, the implementation of 5G technology in Seoul and Busan through smart city projects in South Korea transformed telemedicine services. According to Kuzior and Kuzior (2020), it provided intelligent traffic systems while introducing AI-driven security features. Developing countries face challenges in 5G deployment because of high expenses, insufficient supporting infrastructure, and weak government cooperation efforts. The substantial advantages of 5G and edge computing face the challenge of narrowing the global urban technology gap, which is exacerbated by varying levels of preparedness between regions (Calzada 2020).

Again, Startups and technology hubs operate as drivers of smart city innovation because they deliver disruptive technologies that boost operational excellence, make services more inclusive, and generate economic prospects. The startup scene in Silicon Valley raises billions in investments for enterprises developing AI-powered urban analytics solutions, green energy technology, and smart mobility systems (Sepasgozar et al. 2019). Developing nations face systemic obstacles that hinder the growth of their startups due to inadequate venture capital support, regulatory backing, and foundational infrastructure capabilities (Habib et al. 2023). Smart cities must adopt adaptive and resilient strategies to address rapidly accelerating climate change, resource depletion, and extreme weather events. Hussain (2024) revealed that sustainable infrastructure, AI-based climate monitoring systems, and decentralised renewable power networks will establish fundamental elements for enduring urban sustainability. AI technology operating in Tokyo has lowered disaster-related deaths by more than 30% through its flood prediction system, which notifies citizens about high-risk flood zones (Springs 2024). Smart cities must address three additional critical matters beyond climate change: managing social inequality, protecting cybersecurity, and addressing data protection concerns. Belaid et al. (2024) indicate that the use of artificial intelligence for authority surveillance in China and India has sparked public discussions about citizens' privacy rights and government control issues, highlighting the ethics of digital urbanisation. The future will face a crucial challenge in maintaining smart cities that are inclusive, democratic, and secure for all residents.

4.7 Conclusion

This chapter examined the dynamics of smart cities in both developed and developing economies, highlighting the distinct development and convergence as they progress in their urban transformation. In developed economies, the responsibilities of a smart city are typically facilitated by the existence of robust digital infrastructure, the integration of AI in governance processes, and high-speed connectivity, which ensures greater efficiency and resilience within urban systems. Conversely, emerging markets tend to operate on mobile-first innovations, outside financing, and minimalistic technological answers to infrastructural shortcomings and social inequality.

Nevertheless, these structural differences do not compare with the list of shared problems, which in both cases include digital inequality, cybersecurity risks, and issues of data privacy. The resurgence of these concerns reinforces the importance of an all-inclusive approach to digital strategies. The issue of digital inclusion, which refers to equal access to and the appropriate utilisation of digital technologies, has become a primary concern. The leaders will need to introduce educational initiatives that promote digital literacy and make smart services accessible to vulnerable populations, including low-income families and older citizens.

Moreover, another concept which is central to funding and implementing smart city technologies is public-private partnerships (PPPs). Nevertheless, this growing concentration of insiders in the field evokes questions of monopsonization and inequitable access to vital services. Governments are obligated to establish a proper system of regulations to maintain accountability, transparency, and the welfare of the people in such collaborations. Cybersecurity and data privacy are also becoming essential aspects of smart city governance. As the adoption or urban management of AI and IoT increases, cities need to implement robust data protection regulations and ethical AI governance that can protect the rights of citizens and ensure that areas of surveillance do not expand inappropriately. The development of the infrastructure should consequently be done in harmony with social responsibility-integrating smart technologies with sustainable mobility, effective waste management, and inclusive digital platforms.

The chapter also enumerates the benefits of international collaborations in overcoming the technological gap between developed and developing nations. It is agreed that international organisations, including the World Bank and UN-Habitat, will assist in cross-border knowledge transfer, capacity development, and funding of smart infrastructure projects. This type of partnership has the potential to introduce locally tailored solutions and promote sustainable urban development on a global scale.

Consequently, effective smart city adoption entails a delicate strategic combination of technological and social progress supported by architectural integrity and clear governance, as well as environmental sustainability. The future of smart cities lies in their ability to integrate digital innovations, ethical standards, participatory planning, and resilient resource management systems. It is only under such comprehensive measures that cities can be transformed into developing equitable, sustainable, and livable communities that accommodate all their citizens.

References

Addas A (2023) The concept of smart cities: a sustainability aspect for future urban development based on different cities. Front Environ Sci 11:1241593

Angelidou M, Psaltoglou A, Komninos N, Kakderi C, Tsarchopoulos P, Panori A (2018) Enhancing sustainable urban development through smart city applications. J Sci Technol Policy Manage 9(2):146–169

Artmann M, Kohler M, Meinel G, Gan J, Ioja IC (2019) How smart growth and green infrastructure can mutually support each other—a conceptual framework for compact and green cities. Ecol Indic 96:10–22

Attaran H, Kheibari N, Bahrepour D (2022) Toward integrated smart city: a new model for implementation and design challenges. GeoJournal 87(Suppl 4):511–526

Aurigi A, Odendaal N (2022) From "smart in the box" to "smart in the city": rethinking the socially sustainable smart city in context. In: Sustainable smart city transitions. Routledge, pp 53–68

Bai L, Xie L, Li C, Yuan S, Niu D, Wang T, Yang Z, Zhang Y (2023) The conceptual framework of smart TOD: an integration of Smart City and TOD. Land 12(3):664

Bandauko E, Nutifafa Arku R (2023) A critical analysis of 'smart cities' as an urban development strategy in Africa. Int Plan Stud 28(1):69–86

Bauer M, Sanchez L, Song J (2021) IoT-enabled smart cities: evolution and outlook. Sensors 21(13):4511

Belaid F, Ramdani B, Abdellatif M (2024) Can Smart Cities Be Sustainable? An Emerging Field of Research. J Urban Technol:1–22

Bellini P, Nesi P, Pantaleo G (2022) IoT-enabled smart cities: a review of concepts, frameworks and key technologies. Appl Sci 12(3):1607

Caird SP, Hallett SH (2019) Towards evaluation design for smart city development. J Urban Des 24(2):188–209

Calzada I (2020) Democratizing smart cities? Penta-helix multi-stakeholder social innovation framework. Smart Cities 3(4):1145–1172

Chen J (2023a) Mitigating nitrogen dioxide air pollution: the roles and effect of national smart city pilots in China. Energy 263:125652

Chen P (2023b) Urban planning policy and clean energy development Harmony-evidence from smart city pilot policy in China. Renew Energy 210:251–257

Davis FD (1989) Technology acceptance model: TAM. Al-Suqri, MN, Al-Aufi, AS: Information Seeking Behavior and Technology Adoption 205(219):5

De Falco S (2019) Are smart cities global cities? A European perspective. Eur Planning Stud 27(4):759–783

De Falco S, Angelidou M, Addie JPD (2019) From the "smart city" to the "smart metropolis"? Building resilience in the urban periphery. Eur Urban Reg Stud 26(2):205–223

Duygan M, Fischer M, Pärli R, Ingold K (2022) Where do Smart Cities grow? The spatial and socioeconomic configurations of smart city development. Sustain Cities Soc 77:103578

Filiou D, Kesidou E, Wu L (2023) Are smart cities green? The role of environmental and digital policies for Eco-innovation in China. World Dev 165:106212

Guo Q, Wang Y, Dong X (2022) Effects of smart city construction on energy saving and CO2 emission reduction: evidence from China. Appl Energy 313:118879

Habib A, Alsmadi D, Prybutok VR (2023) Factors that determine residents' acceptance of smart city technologies. In: Smart Cities at play: technology and emerging forms of playfulness. CRC Press, pp 4–17

Hatuka T, Rosen-Zvi I, Birnhack M, Toch E, Zur H (2018) The political premises of contemporary urban concepts: the global city, the sustainable city, the resilient city, the creative city, and the smart city. Plan Theory Pract 19(2):160–179

Hou J, Arpan L, Wu Y, Feiock R, Ozguven E, Arghandeh R (2020) The road toward smart cities: a study of citizens' acceptance of mobile applications for city services. Energies 13(10):2496

Hussain I (2024) Secure, sustainable smart cities and the internet of things: perspectives, challenges, and future directions. Sustainability 16(4):1390

Iryna K, Iryna U, Yevhen P, Nataliia D, Maxym B (2022) Factors of successful development of smart cities. Int J Comput Sci Network Security 22(7):21–28

Ismagilova E, Hughes L, Dwivedi YK, Raman KR (2019) Smart cities: advances in research—an information systems perspective. Int J Inf Manag 47:88–100

Jonek-Kowalska I, Wolniak R (2021) Economic opportunities for creating smart cities in Poland. Does wealth matter? Cities 114:103222

Jonek-Kowalska I, Wolniak R (2022) Sharing economies' initiatives in municipal authorities' perspective: research evidence from Poland in the context of smart cities' development. Sustainability 14(4):2064

Khan HH, Malik MN, Zafar R, Goni FA, Chofreh AG, Klemeš JJ, Alotaibi Y (2020) Challenges for sustainable smart city development: a conceptual framework. Sustain Dev 28(5):1507–1518

Kim J (2022) Smart city trends: a focus on 5 countries and 15 companies. Cities 123:103551

Kirimtat A, Krejcar O, Kertesz A, Tasgetiren MF (2020) Future trends and current state of smart city concepts: a survey. IEEE Access 8:86448–86467
Komninos N, Kakderi C, Panori A, Tsarchopoulos P (2019) Smart city planning from an evolutionary perspective. J Urban Technol 26(2):3–20
Kummitha RKR, Crutzen N (2017) How do we understand smart cities? An evolutionary perspective. Cities 67:43–52
Kuzior A, Kuzior P (2020) The quadruple helix model as a smart city design principle. Virtual Econ 3(1):39–57
Masik G, Sagan I, Scott JW (2021) Smart City strategies and new urban development policies in the Polish context. Cities 108:102970
Matos F, Vairinhos V, Durst S, Dameri RP (2019) Intellectual capital and innovation for sustainable smart cities: the case of N-Tuple of helices. In: Intellectual capital management as a driver of sustainability: perspectives for organizations and society. Springer, pp 49–66
Mazzetto S (2024) A review of urban digital twins integration, challenges, and future directions in smart city development. Sustainability 16(19):8337
Moura F, de Abreu e Silva J (2021) Smart cities: definitions, evolution of the concept, and examples of initiatives. In: Industry, innovation and infrastructure. Springer International Publishing, Cham, pp 989–997
Okafor CC, Aigbavboa C, Thwala WD (2023) A bibliometric evaluation and critical review of the smart city concept–making a case for social equity. J Sci Technol Policy Manage 14(3):487–510
Qayyum S, Ullah F, Al-Turjman F, Mojtahedi M (2021) Managing smart cities through six sigma DMADICV method: a review-based conceptual framework. Sustain Cities Soc 72:103022
Qonita M, Giyarsih SR (2023) Smart city assessment using the Boyd Cohen smart city wheel in Salatiga, Indonesia. GeoJournal 88(1):479–492
Rifaid R, Abdurrahman A, Baharuddin T, Kusuma BMA (2023) Smart city development in the new Capital City: indonesian government plans. J Contemp Gov Public Policy 4(2):115–130
Ristvej J, Lacinák M, Ondrejka R (2020) On smart city and safe city concepts. Mob Netw Appl 25:836–845
Ruggerio CA (2021) Sustainability and sustainable development: a review of principles and definitions. Sci Total Environ 786:147481
Sepasgozar SM, Hawken S, Sargolzaei S, Foroozanfa M (2019) Implementing citizen centric technology in developing smart cities: a model for predicting the acceptance of urban technologies. Technol Forecast Soc Chang 142:105–116
Sharifi A (2019) A critical review of selected smart city assessment tools and indicator sets. J Clean Prod 233:1269–1283
Sharma S, Popli R, Singh S, Chhabra G, Saini GS, Singh M, Sandhu A, Sharma A, Kumar R (2024) The role of 6G technologies in advancing smart city applications: opportunities and challenges. Sustainability 16(16):7039
Siemens (2021). Sustainability report 2021. https://assets.new.siemens.com/siemens/assets/api/uuid:4806da09-01c7-40b1-af91-99af4b726653/sustainability2021-en.pdf
Shu Y, Deng N, Wu Y, Bao S, Bie A (2023) Urban governance and sustainable development: the effect of smart city on carbon emission in China. Technol Forecast Soc Chang 193:122643
Springs D (2024) Elements of Smart Leadership Approaches for Smart City Development. Business Ethics Leadership 8(2):35–48
Statista (2025) Share of urban population worldwide in 2023, by continent. https://www.statista.com/statistics/270860/urbanization-by-continent/
Suárez-Eiroa B, Fernández E, Méndez-Martínez G, Soto-Oñate D (2019) Operational principles of circular economy for sustainable development: linking theory and practice. J Clean Prod 214:952–961
Tan SY, Taeihagh A (2020) Smart city governance in developing countries: a systematic literature review. Sustainability 12(3):899
Ullah A, Qi G, Hussain S, Ullah I, Ali Z (2024) The role of LLMs in sustainable smart cities: applications, challenges, and future directions. arXiv preprint arXiv:2402.14596
Urbanet (2025) The world urban population | Infographics. https://www.urbanet.info/world-urban-population/#:~:text=According%20to%20projections%2C%20the%20urban,population%20in%20sub%2DSaharan%20Africa. Accessed 10 Feb 2025
Waqar A, Othman I, Almujibah H, Khan MB, Alotaibi S, Elhassan AA (2023) Factors influencing adoption of digital twin advanced technologies for smart city development: evidence from Malaysia. Buildings 13(3):775
Wu D, Xie Y, Lyu S (2023) Disentangling the complex impacts of urban digital transformation and environmental pollution: evidence from smart city pilots in China. Sustain Cities Soc 88:104266
Yadav G, Mangla SK, Luthra S, Rai DP (2019) Developing a sustainable smart city framework for developing economies: an Indian context. Sustain Cities Soc 47:101462
Zheng C, Yuan J, Zhu L, Zhang Y, Shao Q (2020) From digital to sustainable: A scientometric review of smart city literature between 1990 and 2019. Journal of Cleaner Production 258:120689

5 Migration, Population Growth, and Threats for Smart Urban Cities

Abstract

Rapid population growth has a negative social and economic impact on the urban city. Population growth can be the result of increasing migration from rural to urban cities. It reduces the quality of life, reduces welfare level, endanger safety and security of urban city. It is a threat for the resilience of the city. Additionally, it causes pollution, lack of utility services, and the source of congestion. The housing market becomes vulnerable, and rent gets more expensive for everyone. Buying new houses in highly populated urban cities is also difficult due to increasing demand. The price of new dwellings is almost double, even triple in crowded urban cities because of supply constraint demand. To resolve such utility problems and increase energy efficiency, smart cities must install smart meters. Moreover, installing smart water meters and accessing well developed information computer technology (ICT) are essential in smart urban cities. Readers of this chapter will have comprehensive information about the needs of a smart city during the transformation process, especially as a case study of Catalkoy-Esentepe region. Urban mobility causes an increasing number of motor vehicles in use, which causes severe traffic congestion, carbon emission, noise pollution, hazards of road accidents, and waste of financial resources on gasoline expenditures. To safeguard carbon free environment, it is proposed to expand bicycle roads and pedestrian paths. Public transport must be improved in Catalkoy-Esentepe region with newly built ring roads, bypasses, underpasses, elevated roads and installation of digital signalization on crossroads. This chapter will explore the discussions about the concepts of rainwater management, sanitation, solid waste management, energy efficiency, rehabilitation of green and blue spaces, construction of advanced technology of internet infrastructure, good governance and the need for spatial planning for the development of Catalkoy-Esentepe as a case study of future smart city in North Cyprus.

5.1 Introduction

Urban mobility from rural areas to urban cities causes population growth and creates social and economic problems. The increasing population, municipal services become inadequate, and causing inadequate utility services, lack of water, electricity, telephone, health, education and other similar facilities, which are provided publicly to all residents. Residential areas also suffer from traffic congestion, waste and noise pollution. Additionally, housing demand increases, and rent

V. Yorucu et al., *Modern Approach of Resilient and Sustainable Smart Cities*, Sustainable Development Goals Series, https://doi.org/10.1007/978-3-032-05715-0_5

becomes more expensive. High rent provides an opportunity for contractors to develop mass housing projects, which limits the availability of use of land. According to Amine (2023), addressing the challenges of high population growth and urbanization, urban agglomerations need to be designed intelligently, enabling sustainable living in smart cities (Belaïd and Arora 2024). The meaning of urbanization relies on driving forces, such as mobility, entrepreneurial dynamics, efficiency of energy management, transport, IT system etc. Chatterjee and Kar (2015) have noted that the pivotal role for the conversion of developed cities into standard smart cities includes metropolistic competition and product variety. The conversion is a continuous process of improvement by the help involvement of citizens, use of hard infrastructure, use of modern technologies which would ultimately make the cities livable, resilient and capable of combatting challenges. One of the advantages of using information computer technology is to increase efficiencies, minimize costs and to enhance quality of life as well as the welfare level of people. The citizens of a smart city must be well educated, be aware of using technology, have easy access to all information they need to make choices about their daily lifestyle work and travel options. Moreover, in this chapter, readers can have detailed information on urban development, urban mobility, housing, green and open spaces, water management, blue spaces and disaster risk management, waste management, urban policy and governance.

5.2 Literature Survey on Migration and Population Growth

We have learned from the related literature that many countries have faced similar problems, such as Saudi Arabia, India, China, Turkey and many more. Chen et al. (2023) pointed out that smart cities are more favorable destinations for distant migrants, and migrants with higher educational attainment and income. Such argument was based on the evidence that in China, migrant survey data was implemented for a research study and the findings highlight the importance of digitalization in urban development.

There are many smart cities available in the world who successfully completed the transformation process. These cities use technology and innovation to solve urban problems. Nordic countries top the list, Stockholm, Sweden rating an A+, the only city to do so. Of the top ten sustainable cities, no less than five are Nordic countries (Sweden, Norway, Denmark, and Finland), all having a population of less than one million. Two cities are from Canada (Vancouver and Toronto), plus more megacities, London, UK, Tokyo, Japan and Berlin, Germany (Mehmet and Yorucu 2024).

These are model cities whose experience may serve as guidelines for the rest of the urban centers in the world. All these cities merit a rating of A+ in Air Quality and Climate Change Resilience, which are compliant with SDG 13. This strongly suggests that in the most sustainable cities, municipal action to protect the environment is input from below rather than central government policy from the top. Citizens in neighborhoods must first acknowledge the necessity of environmental protection, demanding sustainable policies at the city level. National action for sustainability typically follows city-level policies.

With respect to the oil-rich countries, such as Saudi Arabia and Dubai in the Middle East, it is important to sound a word of caution. As the age of fossil-fuel draws to a close, the long-term sustainability of these states becomes questionable. It is possible, of course, that with smart financial planning, the long run, even a few centuries into the future, may be secured. No society, however, can live off interest and dividend payments forever, unless something like the Singapore model is created as a Smart City with a world-class financial center.

Mehmet and Yorucu (2024) stated that in Singapore, commercial enterprises have been improved through the publication of real-time data on the operation of city services. The latest progress in technology has helped to make banking and financial innovation virtually limitless.

Beyond becoming a world-leading financial hub, Singapore has managed to use cutting-edge artificial intelligence (AI) with digital sensors and digital control systems that have helped to control traffic, to create building management systems, digital electricity, gas, water meter and other utilities without human power interference and given a chance to ease congestion, reduce city pollution and waste of human and non-human economic resources. Such an experience of successful transformation process of Singapore is aimed to be implied in our case study of Catalkoy-Esentepe in this monograph.

Singapore is arguably the most successful Smart City. Human capital formation in universities and technology centers matches the best in the world. Shared prosperity not only enables high income per capita; it generates social capital with harmony across its multi-cultural population. Widely used worldwide wireless network, smart phones, and latest information management techniques, have all contributed to Singapore's success story.

More specifically, Anwar and Oakil (2024) stated that Saudia Arabia in the Middle East has faced significant urbanization and population growth that have incentivized the need to establish smart cities. Anwar and Oakil (2024) have noted that during the last four decades, the population of Saudi Arabia has tripled and reached from 9.32 million to 26.3 million. They emphasized that this high population growth alongside migration from rural to urban areas induced urgency in devoting attention to urban development in Saudi Arabia. The most common problems they face in Saudi Arabia is the exhaustion of natural resources, such as water, land, and minerals, and other social problems such as road congestion and pollution, which calls special attention to implement new solutions to address these challenges. Rodrik (2018) pointed out that the larger the population, the greater the number of people who find their preferences ill served by the specific public good that is provided. Rodrik (2011) highlighted that during the last couple of decades capital flows boomed and most of the world's economies became financially more integrated than ever before. He noticed that this was an era of vast flows of people between continents, with working-class Europeans migrating in a group to the Americans and other lands of recent settlement. This has been considered as the first era of globalization. Rodrik (2011) stated that the world has still to catch up to control the problem of labor mobility. Krugman (1992) pointed out that every modern nation has restrictions on labor mobility. Many nations place restrictions on the movement of capital or at least threaten to do so. Krugman (1992) has also noted that urban population growth is the result of labor mobility from peripheries to the core locations. Lower transaction costs in the manufacturing industry have attracted many low-skilled laborers to move from the agricultural industry to the manufacturing industry with the expectation to have better jobs with higher income in manufacturing industries. This has increased the growth of population in urban cities, which has threatened climate change as well as public utilities. In some cities wealth is distributed unevenly due to in-migration or out-migration. Butsch et al. (2017) stated that the population development and the increasing socioeconomic polarization are linked to the economic development of the city. In many cities, the informal economy is the most relevant source of income for a large part of the population. Considering Northern Cyprus, economic development is challenged by informality, poor infrastructure and inadequate planning and governance. Likewise, the region of Catalkoy-Esentepe has grown very rapidly over the last three decades, especially the rapid growth in the construction sector which has the biggest share of the informal sector. A special challenge for urban planning is the transformation of urban fringe areas of the region (including the city of Kyrenia, which is the tourist resort), as this process is currently taking place in an unsustainable manner. Stiglitz (2013, 2015) noted that the informal economy creates income inequality. Butsch et al. (2017) underlined that wealth, which is distributed unevenly, causes social disparities among classes and caste groups. Winters (2011) stated that wages in highly educated cities are higher than in less educated cities, even after controlling individual worker characteristics.

Bhagwati and Panagariya (2013) stated that urban mobility and rapid population growth in India had a negative impact on growth and caused poverty. Bhagwati (1988) argued that growth would create more jobs (in the rural sector itself) and opportunities for gainful improvement in income (as, for example, through migration to growing urban areas), directly pulling more of the poor above the poverty line and additionally would allow the government to pull in more revenues, which would enable the government to spend more on healthcare, primary and secondary education and other development programs to support the poor people. Healthcare, education, and other programs to further help the poor need serious reforms to resolve the poverty problem and sustain economic growth. They argued that if growth did follow liberal policies and reforms in the Indian economy, still some critics argue that the growth is not inclusive, that it has failed to reduce poverty and has not spread to the marginalized groups in society. To create inclusive, safe, resilient, sustainable and smart urban cities, which is one of the UN's sustainable development goals (SDG 11), a nation must plan its population growth and control urban mobility.

Bhatt and Roychoudhury (2023) argued that the transformation of Indian economy caused the share of service sector in India to increase, with a falling share of agriculture, and migration towards cities would continue to grow exponentially. This is the result of increasing economic migration, which resulted in the growth of informal housing, unplanned expansions, traffic congestion, and the deterioration of law and order (Sharifi and Salehi 2022). The increased migration in India has also increased the pressure on the city's civic infrastructure, leading to a severe decline in service quality. Bhatt and Roychoudhury (2023) pointed out that the increased migration has also increased inequality as the rural poor have opted to migrate to urban areas in search of economic opportunities. Urbanization needs to be planned to prevent migration in a way that policy initiatives should focus on the development of a city which connects to the existing civic infrastructure. Landman (2021) emphasized that increasing population growth in rural areas due to migration in many cases has increased levels of inequality, poverty, food insecurity, and unemployment, while warnings about the impact of climate change continue to raise concerns. Acemoglu and Robinson (2012) argued that the great inequality of the modern world that emerged in the nineteenth century was caused by the uneven dissemination of industrial technologies and manufacturing production. Urban mobility is therefore inevitable from rural to urban areas because foreign direct investment in industrial technologies is mostly in urban cities where there is no lack of infrastructure. Acemoglu and Johnson (2023) noted that the contemporary environmental movement confronting the existential threat of climate change demonstrates that the formula's three prongs can work to redirect technological change today. They highlighted that despite the continued reliance of most large energy companies on fossil fuels; there is evidence of noticeable transformation in renewable energy technologies. Rodrik (2011) emphasized that industrialization depended on importing skills and institutions. Intercontinental labor mobility was a tremendous advantage here. He added that, "where Europeans settled en masse, they brought with them both the skills and the drive for more representative, market-friendly institutions that would promote economic activity alongside their interest. The consequences were disastrous for the native populations, who perished in large numbers courtesy of European aggression and germs", (Rodrik 2011, p. 139).

5.3 The Population Growth in Catalkoy-Esentepe Municipal Region During 1831–2024

The population size of the municipality has changed over the period of almost two centuries. Since 1831, the population size for the municipal administration is still estimated to be less than 15,000 inhabitants. Some argue that after the construction boom in the region, the de-jure population has already exceeded 20,000. The figures after the 2011 population census are the projections. More precise figures can be obtained after a new population census in 2025.

As can be seen from Fig. 5.1 above, the population of Catalkoy-Esentepe municipality during 1831–2024 period grew significantly. This figure simply elaborates the trend ov population among six villages over the last two centuries. It will shed light on the readers to understand how population growth in specific villages is fast, whereas some others turned out to be deserted villages. During the Ottoman administration, there was a population census in 1831 and the total number of people living in the villages was 211. Those six villages are composed of today's municipality of Catalkoy-Esentepe with the local reform in 2022. The government has decided to unite some of the inefficient municipalities with the prosperous ones after enacting new legislation which was voted and entered into force in 2022 by the TRNC parliament. Then the total number of municipalities reduced from 28 to 18. During the British Administration in Cyprus (1878–1958), British colonial government has undertaken six population censuses between 1891 and 1946.

During that period the total aggregated population for the region of Catalkoy-Esentepe ranged from 1698 to 4146. The population of Besparmak, Karaagac and Bahceli were not significant at all and the population trend for such villages was stationary. When the Republic of Cyprus was established in 1960, the population census was repeated, and the total population of the whole area was reported as 4146. From 1960 to 1973, the population for the region stayed almost the same and reported as 4195. Nonetheless, during the trouble period of the Republic, the Turkish Cypriots living in Besparmak village was halved from 132 to 62. There was not any Greek Cypriots living in Besparmak after 1921. Only a few Greek Cypriot shepherds were recorded in 1905 (only 6 habitants) and 1911 (15 habitants), respectively. The highest population record ever Besparmak had was 132 in 1973, and then the number of inhabitants living in the village reduced continuously to the record low (nearly 30 inhabitants), back to the level of early 1900. This village needs to be protected and has potential advantages to be transformed into a Citta slow village. The village is very similar to the abandoned village of Variseia, in Trodos, near Alevka. Considering Bahceli, there was no influence from humans, therefore the habitat and plants have flourished in the village. The animals, vegetation, and many more are growing in harmony and shining with smiles. We cannot make any similar comments to

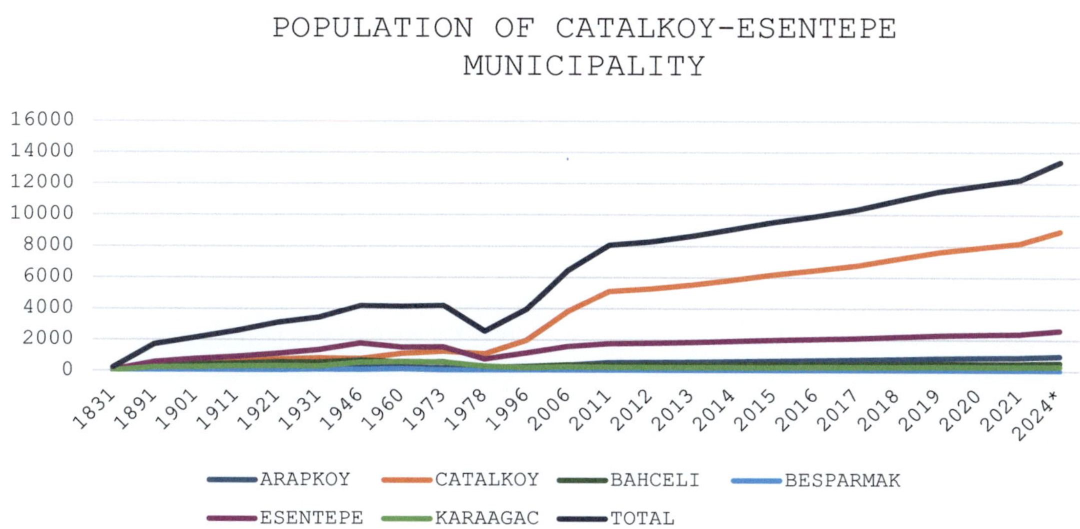

Fig. 5.1 The population growth of Catalkoy-Esentepe Municipal Region. (Source: Cyprus Turkish Statistics Association, Nicosia, Prime Ministry Office, 2024. Authors' compilation)

Catalkoy and Esentepe villages due to unplanned city planning and expansion in population. There was a structural break in total population growth during 1978 population census due to the war that took place in 1974 and Cyprus was partitioned in two. The Greek Cypriot population migrated to the South and Turkish Cypriot to the North. Following the mobility policy of Turkish Cypriot Administration in 1975, most villages population were halved. To make the farming and agriculture continuous, Turkish Cypriot administration decided to accept migrant workers and families from Turkey. People from the Black Sea region of Turkey, mostly from Trabzon and Rize have moved to Esentepe and Bahceli, and some from other regions of Turkey settled down in Catalkoy and Karaagac. Following economic growth, construction boom and expansion in the tourism sector, the population of the Catalkoy-Esentepe region is projected to be 13,385 (see Table 5.1 below). Some argue that de-jure population estimated figures do not reflect the actual reality. Some indicators, such as number of tourists staying in the tourist accommodation, number of students in the university and number of water meters installed by the municipality, number of cars running in the region, amount of bread sold by the markets daily, and the number of subscribers using mobile telephones indicate much bigger population projections. Table 5.2 includes population data from different sources which tabulates the statistics of de-jure population over the period 1831–2024. The Fig. 5.1 illustrated above shows the movements of population trend over the last two centuries. The declining trend during 1974 is the result of the war between two communities and migration of refuges, as also called displacement in Cyprus problem. Greek Cypriots who were living in the North forced to move to the South, and Turkish Cypriots to the North. As bi-communality and bi-zonality have been accepted by two communities with the UN supported resolutions, the population statistics in the villages located in the North declined during 1974–1978. Turkish Cypriot Administration has implemented new housing policy and accepted Turkish immigrants from Turkey to come and settle in the villages listed below in Table 5.1. This was the result of economic inevitability during the post-war period. Declined population, followed with new refugee policy from the Southern part of Cyprus and migration policy of the Turkish Cypriot Administration to accept Turkish settlers from Turkey ended up creating new dispossessed owners in 1978. This has created a property dispute among Cypriots and no solutions have been found so far over the past 51 years.

5.4 Urban Development Priorities and Challenges of Catalkoy-Esentepe

As population increase threatens the city of Catalkoy-Esentepe, there is an urgent need to investigate the other issues of urban sprawl. The city calls special need to assure resilience. In this section, the main aim is to underline the main urban development challenges under each of the priority areas. The priorities can be summarized as urban mobility, housing, green and open spaces, water management, bule spaces, disaster risk management, waste management, urban policy and governance framework. Each section will have sub-sections to support the explanations, arguments and suggestions.

5.4.1 Urban Mobility

Hachette and L'Hostis (2024) noted that in recent decades, mobility policies and transportation services have been evolving rapidly, mainly in large and medium-sized cities. Moreover, they stated that the aim of public authorities is to meet the targeted objective of sustainable mobility. The city of Catalkoy-Esentepe has a sort of car-centric mobility system, which reflects the lack of adequate public transport services. There are minibus services running every hour from 7:30 am to 6:00 pm, which is not suitable for someone who wants to socialize and chooses to go out and decides to stay until very late in the evening. There is no public transportation to travel after 6:00 pm, and the only option is to take expensive

Table 5.1 The population statistics of Catalkoy-Esentepe Municipal Region during 1831–2024

	Population						
Years	Arapkoy	Catalkoy	Bahceli	Besparmak	Esentepe	Karaagac	Total
1831	20	64	12	22	70	23	211
1891	129	398	352	21	566	232	1698
1901	170	444	431	71	764	243	2123
1911	238	594	471	53	906	297	2559
1921	283	753	566	63	1114	318	3097
1931	268	827	553	107	1360	314	3429
1946	328	777	678	79	1778	541	4181
1960	233	1108	571	132	1508	594	4146
1973	201	1252	578	62	1543	559	4195
1978	145	1091	226	30	757	300	2549
1996	268	1961	290	42	1153	222	3936
2006	380	3803	393	30	1575	264	6445
2011	542	5110	388	30	1754	242	8066
2012	559	5268	393	30	1790	244	8284
2013	589	5541	402	29	1849	248	8658
2014	624	5853	413	29	1918	253	9090
2015	660	6187	425	29	1993	259	9553
2016	691	6466	432	29	2046	262	9926
2017	724	6770	441	29	2105	265	10,334
2018	772	7203	456	29	2201	273	10,934
2019	817	7617	470	28	2288	280	11,500
2020	850	7912	476	28	2336	281	11,883
2021	881	8196	480	31	2379	283	12,250
2024*	963	8956	525	33	2600	309	13,385

Source: Cyprus Turkish Statistics Association, Nicosia, Prime Ministry Office. (2024)
Permission granted by Cyprus Turkish Statistics Association of Prime Ministry, TRNC
Source: Wikipedia. Population statistics for the 1831–1974 period are available at Wikipedia. Accessed on 22 April 2025
https://en.wikipedia.org/wiki/Demographics_of_Cyprus
https://en.wikipedia.org/wiki/Wikipedia:Text_of_the_Creative_Commons_Attribution-ShareAlike_4.0_International_License (Copyright CC BY-SA 4.0 GFDL)
Population for Esentepe (Agios Amvrosios). http://www.prio-cyprus-displacement.net/default_print.asp?id=420
Population for Catalkoy (Agios Epiktitos). http://www.prio-cyprus-displacement.net/default_print.asp?id=422
Population for Karaagac (Charkeia). http://www.prio-cyprus-displacement.net/default_print.asp?id=430
Population for Bahceli (Kalograia). http://www.prio-cyprus-displacement.net/default_print.asp?id=437
Population for Arapkoy (Klepini). http://www.prio-cyprus-displacement.net/default_print.asp?id=445
Population for Besparmak (Trapeza). http://www.prio-cyprus-displacement.net/default_print.asp?id=465

taxis. The Municipality must have public transportation with flexible hours, and the region must convert the fuel source of its obsolete bus fleets to carbon free buses.

Several feasibility studies must be prepared with the support of international organizations, such as the EU Support Office in Cyprus. The city must modernize its main roads. Catalkoy-Girne main road should be expanded to double carriage and the city of Catalkoy needs a ring road to be completed by the national government. The congestion cannot be eased unless the new ring is built urgently. The national physical plan demonstrated below (see Fig. 5.2) displays the new double carriage ring road project. Figure 5.2 is about national highway plan which is extracted from national physical master plan of Northern Cyprus. It shows the direction of the northeastern highway from Kyrenia to Catalkoy-Esentepe. The projected double carriageway and new roundabout construction plans, as well as new ring road from Girne to Çatalköy (namely new

Table 5.2 Statistics of several unexpected climate events in the region of Catalkoy-Esentepe

		Catalkoy			**Esentepe**	**Flood**	05.12.2018	252.5 kg		Four people died		
							01.12.2016	334.2 kg				
	Date			**Date**								
Freezing Cold	20.10.2012	0.0		04.02.1997	−3.5	**Cyclone**	20.11.2020			Industrial zone of Catalkoy		
	10.01.2013	0.0		16.02.2008	−2.0							
	19.02.2015	1.6		20.01.2012	−2.1							
	13.03.2022	1.7		10.01.2015	−2.3	**Hurricane**	2001	118.8 km/h		8.10.2012		119.9 km/h
	15.01.2024	1.0		26.01.2016	−1.5		2004	136.1 km/h		6.01.2019		112.7 km/h
				27–29.01.2017	−1.3		2011	117.4 km/h				
							2021	110.9 km/h				
Heatwave	08.08.1998	39.8		07.07.2000	41.2		2023	136.1 km/h				
	26.06.2007	39.5		10.08.2001	40.0							
	15.07.2012	40.6		28.06.2007	41.9							
	15.7.2023	40.9		01.08.2010	41.6	**Hale**	05.12.2011		25.12.2016		10.11.2024	
				25.06.2019	41.9		17.02.2012		9.12.2017		01.04.2025	
				27.06.2020	41.9		07.01.2013		12.08.2018			
				30.06.2021	41.9		14.01.2014		19.04.2019			
				15.07.2023	39.9		10.04.2015		26.01.2022			
				14.06.2024	41.2							
		No of fire	Donums	No of fire	Donum							
Forest fire	1995 (27 June)	1	59,390.5	1	21,300.5	**Earthquake**						
May-Sept	2015	8	284.0	10	313.5		29.09.1918		6.3	East of Cyprus		
	2016	3	66.1	18	210.5		20.10.1941		6.3	SW (Limassol/Paphos)		
	2017	1	2.0	15	383.0		9.101996		6.5	West (Paphos)		
	2018	6	15.5	14	837.0		28.12.2013		5.5	Antalia		

Source: Firat, A. (2020) Sürdürülebilir Ormancılık Kuzey Kıbrıs (2020). Okman Printing Ltd. Nicosia. Statistics on Forest Fires (Nearly 10 percent of total Forest Area)
Kucukosmanoglu, A. and Ayberk, H. (2003). Kuzey Kıbrıs Türk Cumhuriyeti Büyük Orman Yangınları. Review of the Faculty of Forestry, 53(2), University of İstanbul Press, İstanbul
Meteoroloji Statistics (2025). KKTC Ulaştırma Bakanlığı, Meteoroloji Dairesi, Meteorloji Statistics (2025), Nicosia
Permissions granted by the Meteorology Department of Ministry of Transportation of TRNC and Prof Altay Fırat

Fig. 5.2 Highways master plan and intercity expansion project from Girne to Catalkoy Acapulco Junction. (Source: National Physical Development Plan, Urban Planning Department, Nicosia, (2022). Permission granted by Urban Planning Department of Ministry of Interior Affairs, TRNC)

belt to connect Kyrenia—Catalkoy tourist resort) can be observed in Fig. 5.2.

The Municipality must implement universal design principles and should improve pedestrians' accessibility, reduce traffic congestion and promote active modes of transport such as walking, cycling or using scooters. There are many examples in other successful smart cities which promote bike-share projects, such as Brugge, Copenhagen, and Lund etc. Catalkoy-Esentepe municipality may learn from their experiences. Figure 5.3 shows the Blue-Bike rack in Brugge which is located next to the Brugge railway station. It holds hundreds of brand-new bikes ready to be shared. The Blue-Bike project in Brugge has been found to be one of the most successful projects. There are many more examples implemented by Scandinavian countries, such as Lundahoj Rent and Share Your Bike Project in university town of Lund in Sweeden, Cykeljagt Rent a Bike, Lime, Ridemovi, and Donkey Republic rent a bike project at Copenhagen. Catalkoy-Esentepe municipality can learn from the success story of Belgium, especially from city of Brugge.

The municipality of Catalkoy-Esentepe must have action plans to build bike lanes, which run 7 km to Girne and 23 km to Alagadi and Esentepe, respectively. This will reduce carbon emissions in the region and attract people to use their bike to travel among the villages within the same municipal region. A bikeshare and scooter share demand assessment and system design feasibility study must be developed and intermunicipal cooperation is needed for such a project. E-scooter sharing systems and electric vehicle charging stations infrastructure shall be planned both in Catalkoy and Esentepe. The success story of Barcelona can be used to encourage such superblocks projects to Girne and Catalkoy municipal regions. Zonal parking systems for cars need to be well planned and the payments for the car park should be through a digital mobile application digitally.

Figure 5.4 below shows the Alagadi Natural Park Pedestrian Way and Cycle Road Project. This map has been extracted from the National Physical Development Plan of Urban Planning Department of Northern Cyprus. The map of a 10 km long pedestrian way and cycle road to be

Fig. 5.3 Bicycle share project: blue bike experience from Brugge. (Source: Image created by Vedat Yorucu)

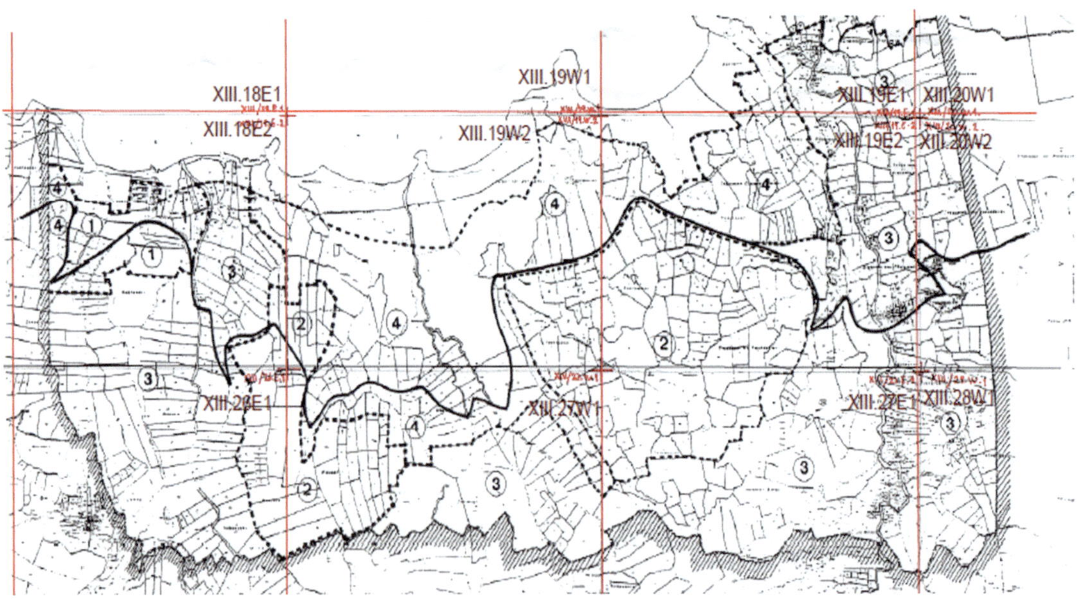

Fig. 5.4 Alagadi Natural Park Pedestrian Way and Cycle Road Project. (Source: National Physical Development Plan, Urban Planning Department, Nicosia, (2022). Permission granted by Urban Planning Department of Ministry of Interior Affairs, TRNC)

constructed in Alagadi Natural Park (coastal region adjacent to Catalkoy) with the earliest period. Once the project is completed, the city of Catalkoy-Esentepe will have a natural park with long pedestrian way facing to the sea in the north and the mountain in the south, which is a fortune to the villagers and tourist to do jogging or enjoyable walks in the nature.

5.4.2 Housing

Old houses in the region of Catalkoy-Esentepe need structural retrofitting and redevelopment. The municipality needs vision and financial capabilities to undertake structural retrofitting and redevelopment. Building used by women for charity, named ÇADER has been revitalized with the funds provided by the municipal sources. With grants, the old Cinema buildings have been transformed into cultural centers both in Catalkoy, and Esentepe. The grants from the rich residents also helped to revitalize those culture centers in both cities. In the main city center of Catalkoy, there is also an old olive oil mill, which has also been revitalized and will be opened for visitors. An old cave, called Sevgi Magrasi in Catalkoy has also been repaired and maintained for cultural visits. Figure 5.5 shown below illustrates the map of Inner City of Catalkoy and the photos exhibits an ongoing revitalization projects of social and cultural centres in Catalkoy and Esentepe.

New Life for the Old Town of Catalkoy and Esentepe, a new vision to organize spatial planning and renovate aged buildings, such as post office, coffee shops, mukhtar's office, and sport club. Repairing public buildings requires both municipal and the local administrations co-financing, especially to repair the adobe houses, barn and stables, building retaining wall alongside the streets, painting front walls and windows of the houses located only in the very central areas, maintaining some roofs, staircases, courtyards, main city hall, wedding hall and likewise.

5.4.3 Green and Open Spaces

Green spaces, woodlands in Kararagac, Alagadi, Bahceli and Arapkoy, forest areas in Catalkoy and Esentepe, agricultural and other unsealed spaces in Arapkoy, and Bahceli were transformed into built-up areas, causing public green area per capita to shrink. The Mayor of Catalkoy-Esentepe has repeated his perseverance to the Ministry of Agriculture and Department of Forestry to open public spaces in the forest area of Esentepe and Catalkoy. The physical development plan showing forest areas is displayed with the map (see Fig. 5.6) below.

Other than green and open spaces, Catalkoy-Esentepe has very serious water and waste management problems. The local administration is very careful about water consumption and scarcity of water during the summer season. As the number of tourists accommodation in the region has increased sharply over the last two decades, water consumption has apparently risen, and local administration needs extra support to prevent leakage and waste. The water sewerage facility nearby tourist accommodations also calls for special attention. Besides, solid waste management and waste of electric electronic equipment must be handled by the municipality. Sections 5.4.4 and 5.4.5 further discuss the topics in detail.

5.4.4 Water Management, Blue Spaces and Disaster Risk Management

The population of Catalkoy extracts water supply from Five-Finger water aquifers. The underground water of the reservoirs of Catalkoy is adequate for only a limited period. The undersea pipeline from Anamur Dragon river of Turkey to Gecitkoy dam in Cyprus has been delivering 75 million cubic meters of water annually and Catalkoy-Esentepe municipal water distribution network has also been connected to the existing water distribution system. Currently, there are no water shortages in the region, yet more water exploration is continuing both in Catalkoy and Esentepe region. There is also spring water running from different parts of the mountain, which satisfy demand for household. Additionally, two big dams in the municipal region, namely Uzundere and Ayandere mostly supply water for agricultural irrigation. There is another small lake in Alagadi region, which has potential to be used whenever there are water shortages. Readers can see Figs. 5.7 and 5.8 below, which exhibit the cadastral map of Uzundere dam in Arapkoy and Ayandere dam in Catalkoy, respectively. Both

Fig. 5.5 Map of Inner City of Catalkoy and ongoing revitalization projects. (Source: National Physical Development Plan, Urban Planning Department, Nicosia, (2022). Permission granted by Urban Planning Department of Ministry of Interior Affairs, TRNC. Photos are the author's compilation)

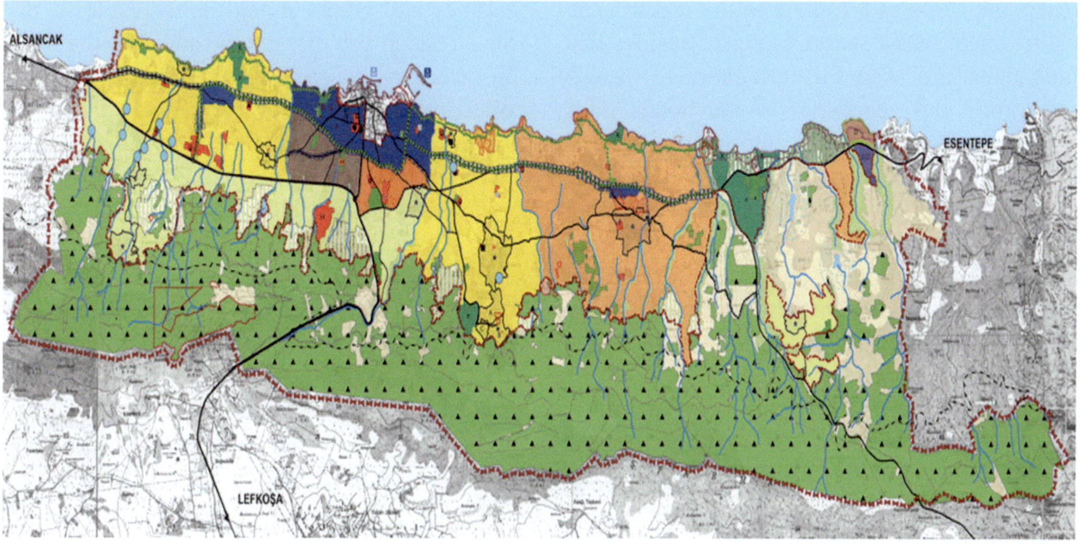

Fig. 5.6 Forest areas of National Physical Development Plan of Catalkoy-Esentepe Municipal Region. (Source: National Physical Development Plan, Urban Planning Department, (2022). Permission granted by Urban Planning Department of Ministry of Interior Affairs, TRNC)

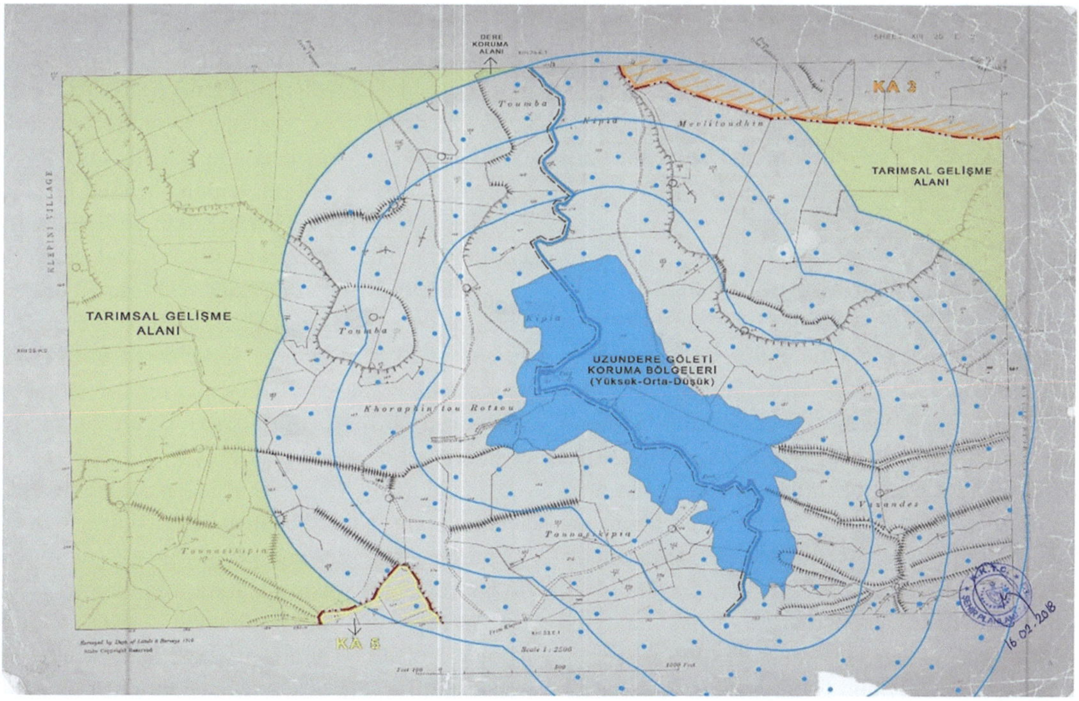

Fig. 5.7 Cadastral map of Uzundere Dam at Arapkoy Village. (Source: National Physical Development Plan, Urban Planning Department, Nicosia, (2022). Permission granted by Urban Planning Department of Ministry of Interior Affairs, TRNC)

reservoirs serve as a lifeline to the region especially when tourist season is peak. Besides that, both dams are used for agriculture to provide fresh fruit and vegetables to the hotels daily. Water supply and wastewater management as integral to climate change mitigation and adaptation as well as the improvement of the city's quality of life, especially since the effects of climate change, like low rainfall events, having drought, are growing burdens on the city. There are seldom flood risks during heavy rainfall due to blockage of some ravines. This impact reflects the vulnerability of Catalkoy to flooding, caused by the city's geomorphology, unplanned urbanization, damped rubble in ravines, sealing of soil surfaces, and malfunctioning in the drainage system. The municipal administration is very active for cleaning main ravines to prevent possible disaster risks, such as soil slide or flood. Additionally, Catalkoy and Esentepe municipal assembly has introduced regulations to prevent the issuance of building permits in risky areas and adopted a flood early warning system. Catalkoy-Esentepe municipality has strengthened the city's resilience to the effects of climate change, including flood risks, and preventing the degradation of surface water quality.

5.4.5 Waste Management

Catalkoy-Esentepe municipality does not have any reliable statistical information about the amount of waste and waste generation is one of the key challenges for waste management in the region. The amount of solid waste collection can be estimated but no one has an idea about the total volume of waste of plastic, paper, metal, electric electronic equipment (WEEE) and glass. The municipality must have a strong focus on recycling its waste. This is not a problem specifically for Catalkoy and Esentepe, but also for the whole Island. To be able to improve waste management in Catalkoy-Esentepe, low interest loan facilities and grants should be available so that low-emission vehicles can be added to the existing inventory to increase the efficiency of waste collection and reduce waste-related emissions. Figure 5.9 shown below illustrates the waste collection containers for recycling. Waste of plastics, paper, glass, cans and metals, as well as electric electronic equipment should be stored separately as a consumer's responsibility. According to the producer's responsibility, collection of waste should also be under care based on the rules and regulations of the EU waste management. Readers can see the Fig. 5.8 Cadastral map of solid waste collection and transfer station in Catalkoy, sewerage plant next to Elexus Hotel, which are adjacent to the Teknecik Power Plant in the village.

Moreover, Municipal Solid Waste Strategy should be developed along with an action plan to bring the city's solid waste management into compliance with the EU requirements. For example, in the South, there are two EU licensed WEEE Recycle Plant and the total amount of minimum WEEE collection must be 10,000 tons/year, however they manage to collect nearly 5000 tons/year. They need the WEEE waste to be collected and stored in depots located in the North and be exported to the South with hook lift containers. The collection, storage and delivery of WEEE must be compliant with the EU rules and regulations. The WEEE, such as air conditioners, and refrigerators have some hazardous gases which causes GHG emissions if it burns and even it may kill people if not properly dismantled.

Additionally, for all kinds of waste, more collection points should be installed in public spaces throughout the city and public awareness needs to be raised for waste separation in the city. The importance of environmental sustainability should be learned at an early age in schools. For reuse and recycle of waste, society must be educated with special programs, such as active lobbying and campaigns.

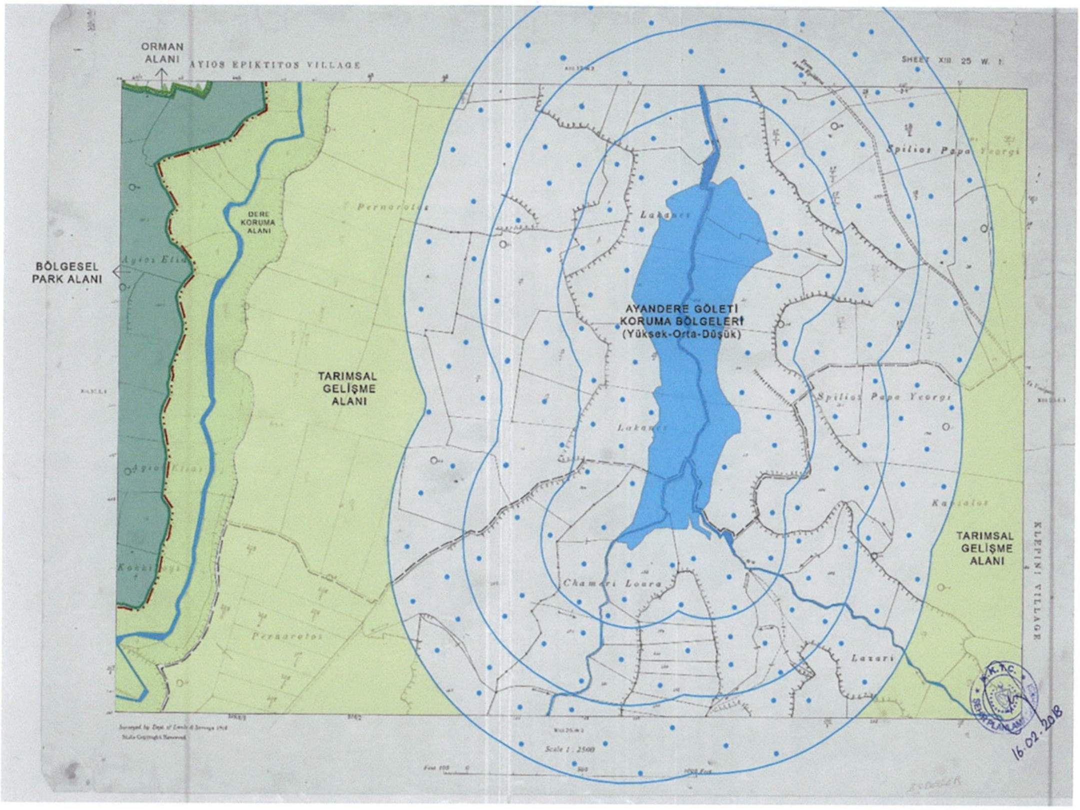

Fig. 5.8 Cadastral map of Ayandere Dam at Catalkoy Village. (Source: National Physical Development Plan, Urban Planning Department, Nicosia, (2022). Permission granted by Urban Planning Department of Ministry of Interior Affairs, TRNC)

5.4.6 Urban Policy and Governance Framework

Chapter 3 delivers more comprehensive information about urban policy and the governance framework. In this chapter, some existing problems can be summarized as unplanned urban sprawl; poor quality housing; shortages of affordable, adequate housing; traffic congestion; air and sound pollution; unfriendly urban design; and lack of green spaces, among others. Considering such problems, Catalkoy-Esentepe should strengthen the capacity of its municipality in the field of spatial and urban planning. The national spatial plan is seeking to expand the territorial coverage of land use plans, such as zoning plans. Such a plan for Catalkoy-Esentepe has new land zoning rules and even strict regulations especially for Alagadi region. Further information regarding such issues will be delivered in the next two chapters.

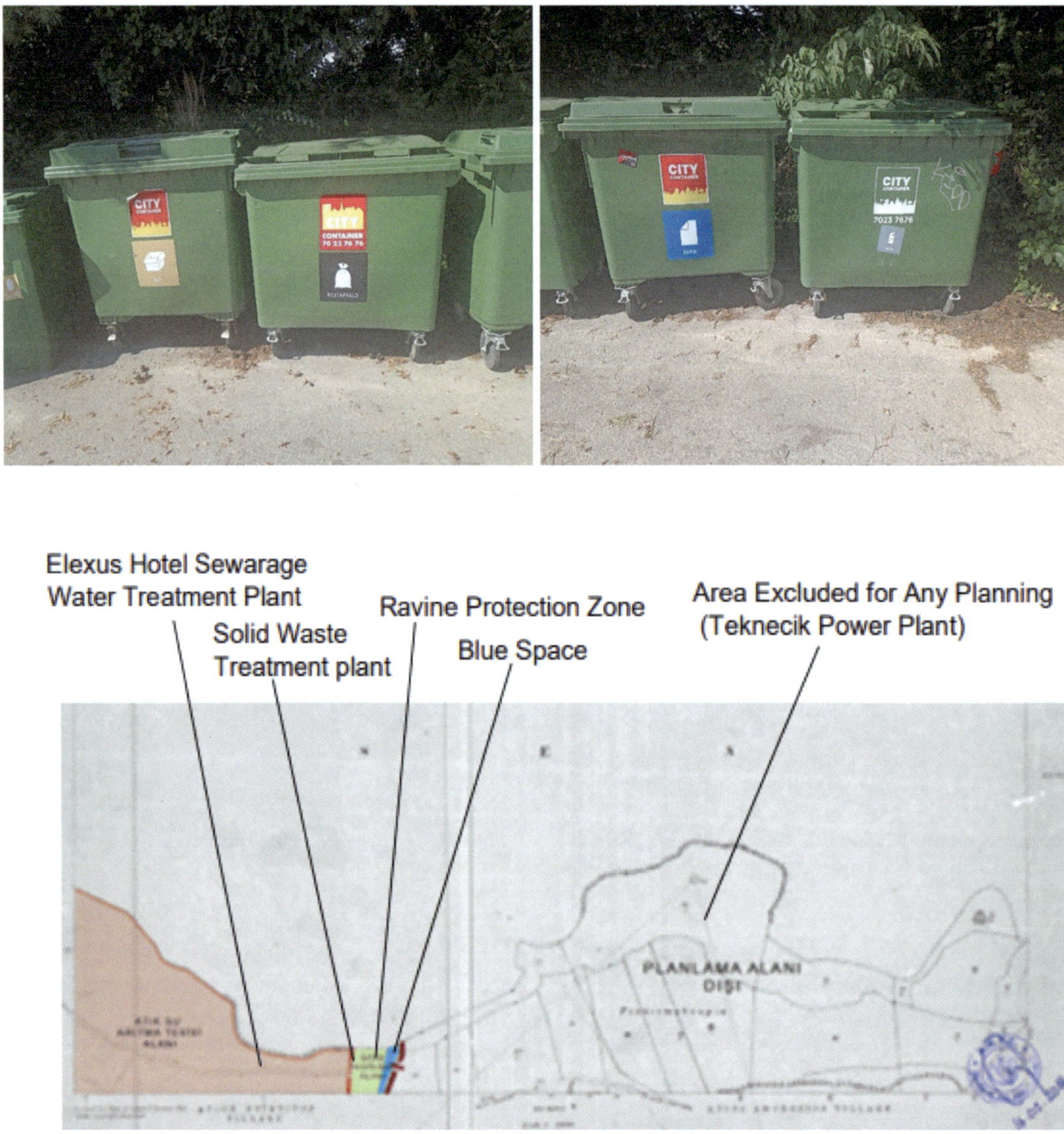

Fig. 5.9 Waste collection containers for recycle and cadastral map for solid waste and wastewater treatment plant in Catalkoy. (Source: National Physical Development Plan, Urban Planning Department, Nicosia, (2022). Permission granted by Urban Planning Department of Ministry of Interior Affairs, TRNC. Photos author's compilation)

5.5 Conclusion

Urban migration is an essential problem of many social and economic problems. In this chapter, we try to focus on rapid population growth and its negative impact on the urban city. We focused more on micro level, analyzing the problems of Catalkoy-Esentepe municipal region, which is currently a major problem, in terms of quality of life, reducing welfare, endangering safety, security and resilience of the city. Besides that, there is serious environmental pollution, as well as insufficient public services provided by the local administrations.

Among many, there are some factors over which attention is required to be focused on the creation of smart cities. For example, considering energy management, Smart Energy Meters should be installed and letting it be used. Demand for energy can then be easily observed; costs can be reduced, and pollution can be controlled. The technology of energy saving including gathering and disseminating data is related to consumption of electricity, therefore control of power sources with effective and interactive ways is essential. Considering electricity production in the region of Catalkoy-Esentepe municipal region, Teknecik Power Plant is endangering human life due to usage of fossil fuel (KIBTEK power company only uses fossil fuels to run turbines) and therefore needs to be substituted with more usage of renewable energy rather than the usage of traditional energy. Cyprus is very ideal location, considering total number of days with plenty of sunshine, and therefore photovoltaic parks will reduce the dependency to Teknecik. Housholds of Catalkoy-Esentepe have already started installing PV panels on their roofs and the number of applicants has increased significantly over the last 5 years. Since energy storage is a serious constraint, households are connected to the grid system. Another alternative solution is to bring electricity from Turkey with sub-sea interconnector gridlines between Cyprus and Turkey. If North Cyprus succeed to be connected to sub-sea electricity interconnector gridlines with Turkey, then the cost of energy will decline significantly, and clean energy will eliminate the existence of carbon emissions caused by Teknecik.

Because of urban mobility, increasing number of motor vehicles in use has given rise to severe traffic congestion, carbon emission and contamination in the air, noise pollution, hazards of increasing road accidents, and waste of financial resources on energy bills. Smart cities should put more emphasis on walkability (Alagadi region of Catalkoy-Esentepe, see Fig. 5.4) and cycling (see Fig. 5.3 Bicycle Share Project) in the city to safeguard carbon free environment. Improving public transport will also help to reduce carbon emissions in urban cities. To multiply urban mobility, the construction of ring roads, bypasses, underpasses, and elevated roads must be completed. Kyrenia municipality is currently improving its road infrastructure in collaboration with Catalkoy-Esentepe local administration. Highways Master Plan and Intercity Expansion Project from Girne to Catalkoy Acapulco Junction (see Fig. 5.2) is under construction and expected to be finalized by 2026. The existing roads are required to be improved; footpaths should be usable by the people walking over those paths making them comfortable to be used.

Water management is also very important in urban smart cities. Smart cities must have enough drinking water as well as water for other use. Every smart city must install smart water meters to measure water consumption by the users. Additionally, smart cities should connect Scada systems to their main water distribution network for examining any possible leakages. An early warning system through Scada network can prevent such water losses. Water contamination may cause an outbreak of epidemics and endanger human life. Ayandare and Uzundere (see Figs. 5.6 and 5.7) water dams in Catalkoy-Esentepe have great potential both for agricultural irrigation and tourism needs. Hotels and tourist accommodation in the region of Catalkoy-Esentepe alone holds almost 50% bedspace capacity among all the 5 stars hotels in Cyprus. Considering total tourists overnight stay, university students, immigrant workers and de jure population, more careful water management is required for Catalkoy-Esentepe.

Sanitation is another issue that needs to be planned and controlled. Lack of proper sanitation would cause an outbreak of epidemic, health hazards and an increase in the mortality rate of the city. The Waste management system in Catalkoy-Esentepe, which has been discussed in this chapter, focuses more on solid waste management system and waste recycling (see Fig. 5.8). It is advisable to search low interest loans as well as grants, either from central administration or from international donors, such as the EU Support Office in Cyprus or UNDP to expand the capacity of low-emission vehicles, constructing container parks, increasing waste collection containers as well as reducing waste-related emissions. Moreover, collection of waste of electric and electronic equipment (WEEE), reuse or exporting it to the Southern Cyprus for recycle is also important for the future of smart city of Catalkoy-Esentepe. The Municipal Solid Waste Strategy should be developed along with an action plan to bring the city's solid waste management into compliance with the EU acquis. Technical knowledge is required to collect, transport and treat solid waste. For proper and scientific waste management systems in smart cities, public awareness for waste collection, separation and recycle is fundamental.

Regarding unexpected climate incidents, sometimes referred as natural disasters, to prevent the impact of forest fire, flood, cyclone, earthquake etc. the municipal administration is on alert to act and implement an early warning system. Installing thermal devices and cameras, cleaning ravines are among some of the actions. Preventing risk of floods has been prioritized and installing cameras and signal detectors to alert the level of rainfall, examining puddles and overflow of rainwater are aimed to be implemented with the help of AI.

Statistics regarding unexpected climate incidents in the municipal region of Catalkoy-Esentepe are displayed in Table 5.2 below.

Considering storm and rainwater drains, Catalkoy-Esentepe municipality has cleaned all ravines running from Five-Finger Mountain through inner city and fall into Mediterranean. All ravines are regularly rehabilitated to prevent floods. In some cities, rainwater drains relate to sewerage distribution networks rendering the system ineffective. Catalkoy municipality has prepared work to preserve rainwater in proper way which will be helpful during summer months when hotels occupancies are high. Ayandere and Uzundere dams are under care and rehabilitated regularly and the water reservoir is well utilized during hot summer and drought.

Energy efficiency in smart cities is another priority objective. The city must install smart meters, and smart grids must be established and integrated with renewable energy sources. Catalkoy-Esentepe households have already installed photovoltaic solar panels and connected them to the grid system. An integrated billing system for electricity, water, telephone, and internet have been set up in a very simple form with the help of an online payment system in the city. However, Teknecik Electricity Plant is running with fuel oil and diesel which endanger human life within a 30-mile radius. The mortality rate in Cyprus is the highest in Catalkoy-Esentepe, mostly caused by cancer. Many protests have been made, yet no action has been taken by the government to substitute oil and diesel turbines with natural gas, in the form of LNG. Alternatively, electricity subsea interconnector grid lines from Turkey could be even better solution to get rid of such an environmental pollution caused by Teknecik.

A well-developed advanced technology internet infrastructure with broadband coverage of the area of cell phones towers should be ensured in Catalkoy-Esentepe, to make the city smart city. The digital transformation of the city and enabling local administration to provide most municipal services electronically as well as good governance through online network requires strong internet technology (5G technology). Smart cities must have reliable and transparent vision and mission with this concern. Data protection and respect for the privacy of data are also other essential elements one should not ignore.

Urban development with essential spatial planning is essential for all smart cities. Catalkoy-Esentepe region in our micro-scale implementation is very successful in spatial planning, which will be discussed in detail in Chaps. 6 and 7. Education, health, entertainment, and sport facilities provide better living standards and increase welfare level of all the citizens in the Smart cities. Catalkoy with leisure activities, nice clean beaches, water sports, Hz Omer recreation area, walking paths, festivities, theaters and sport facilities is a resilient and safe city, which is compliant with the UN SDGs 11 objectives. Newly constructed preschools, and elementary school buildings with the EU building standards in both Catalkoy and Esentepe provide all educational facilities the students and the teachers need.

Social media and access to all information is very important in the concept of smart city. Sharing information and knowledge amongst residents of Catalkoy-Esentepe can be arranged with the connection of social media to make the residents aware of what's going on and be smarter.

The prayer of religious people (Islam or Christianity) must have wider buildings (Church or Mosque), which can accommodate more crowds. Those who pray must enjoy more comfortable spaces to practice their religion. Existing infrastructure capacities for such occasions are not adequate.

References

Acemoglu D, Johnson S (2023) Power and progress: our thousand-year struggle over technology and prosperity. Winners of the 2024 Nobel Prize for Economics, Hachette, UK

Acemoglu D, Robinson JA (2012) Why nations fail: the origins of power, prosperity and poverty. Profile Books Ltd, London, p 51

Amine R (2023) Smart cities: Development and benefits. In Smart cities: Social and environmental challenges and opportunities for local authorities (pp. 45–53). Cham: Springer International Publishing

Anwar AHMM, Oakil AT (2024) Smart transportation systems in Smart Cities: practices, challenges, and opportunities for Saudi cities. Smart Cities 315

Belaïd F, Arora A (2024) Smart cities: social and environmental challenges and opportunities for local authorities. Cham: Springer International Publishing. Springer Nature Switzerland AG: Switzerland

Bhagwati JN (1988) Poverty and public policy. World Dev 16(5):539–555

Bhagwati J, Panagariya A (2013) Why growth matters: how economic growth in India reduced poverty and the lessons for other developing countries. PublicAffairs, New York

Bhatt Y, Roychoudhury J (2023) Smart cities from an Indian perspective: evolving ambitions. In Smart Cities: Social and Environmental Challenges and Opportunities for Local Authorities (pp. 359–383). Cham: Springer International Publishing

Butsch C, Kumar S, Wagner PD, Kroll M, Kantakumar LN, Bharucha E et al (2017) Growing 'smart'? Urbanization processes in Pune urban agglomeration. Sustainability 9(12):2335

Chatterjee S, Kar AK (2015, August) Smart Cities in developing economies: a literature review and policy insights. In: 2015 international conference on advances in computing, communications and informatics (ICACCI). IEEE, pp 2335–2340

Chen X, Cheng M, Yang X, Chu Z, Duan K (2023) Smart cities are more populous: evidence from China. Land 12(10):1917

Firat A (2020) Sürdürülebilir Ormancılık Kuzey Kıbrıs (2020). Orman Yangınları. (Statistics on Forest Fires). Okman Printing Ltd., Nicosia

Hachette M, L'hostis A (2024) Mobility hubs, an innovative concept for sustainable urban mobility?. Smart Cities: Social and Environmental Challenges and Opportunities for Local Authorities. Springer International Publishing, 245–278

Krugman P (1992) Geography and trade. MIT Press, Cambridge

Kucukosmanoglu A, Ayberk H (2003) Kuzey Kıbrıs Türk Cumhuriyeti Büyük Orman Yangınları. Rev Faculty Forestry 53(2)

Landman K (2021) Rapidly changing cities: working with socio-ecological systems to facilitate transformation. Urban Plan 6(2):139–142

Mehmet O, Yorucu V (2024) From land disputes to sustainable environmental development: the near east perspective. Springer Publishers, Switzerland

Meteoroloji Statistics (2025). KKTC Ulaştırma Bakanlığı, Meteoroloji Dairesi, Meteorloji Statistics (2025), Nicosia (Statistics on Climatic Events and Eartquakes)

Plane DA, Jurjevich JR (2009) Ties that no longer bind? The patterns and repercussions of age-articulated migration. Prof Geogr 61(1):4–20

Population for Arapkoy (Klepini) (n.d.). http://www.prio--cyprus-displacement.net/default_print.asp?id=445

Population for Bahceli (Kalograia) (n.d.). http://www.prio-cyprus-displacement.net/default_print.asp?id=437

Population for Besparmak (Trapeza) (n.d.). http://www.prio-cyprus-displacement.net/default_print.asp?id=465

Population for Catalkoy (Agios Epiktitos) (n.d.). http://www.prio-cyprus-displacement.net/default_print.asp?id=422

Population for Esentepe (Agios Amvrosios) (n.d.). http://www.prio-cyprus-displacement.net/default_print.asp?id=420

Population for Karaagac (Charkeia) (n.d.). http://www.prio-cyprus-displacement.net/default_print.asp?id=430

Rodrik D (2011) The Globalization Paradox: democracy and the Future of the World Economy. W.W. Norton and Company Inc., New York

Rodrik D (2018) Straight talk on trade: ideas for a sane world economy. Princeton University Press, Princeton, NJ

Sharifi A, Salehi P (2022) Resilient Smart Cities. The Urban Book Series. doi: https://doi.org/10.1007/978-3-030-95037-8

Stiglitz J (2013) Inequality is holding back the recovery. New York Times 20

Stiglitz JE (2015) The origins of inequality, and policies to contain it. Natl Tax J 68(2):425–448

Winters JV (2011) Why are smart cities growing? Who moves and who stays. J Reg Sci 51(2):253–270

6 The Historical, Cultural and Economic Impacts of Transforming Traditional Urban City to Resilient Smart City

Abstract

This chapter is mainly focused on how a traditional city can be transformed into a safe, inclusive, resilient and sustainable smart urban city, which is the UN's Sustainable Development Goal 11 (SDG 11). The municipal region of Catalkoy-Esentepe has been chosen as a case study to investigate the potential of its transformation process and the necessary action plans, policies and objectives to be implemented to achieve sustainability objectives of SDG 11. The region is very rich with its historical and archaeological sites, that is attracting many foreigners to visit or reside. Additionally, there is a large investment incentive program for the development of capacity building in the tourism sector through new hotels and other types of tourism lodging facilities. Other than tourism, the construction sector boomed in the region during the last decade, and the property market became an important source for foreign exchange income as well as employment opportunities to the region. Such economic developments have triggered urban mobility, which increased population growth, the demand for housing and labor force. Rapid economic growth in the region and urban mobility has negative externalities to city life, such as traffic congestion, increasing rents, insufficient housing, lack of utility services, pollution, climate change and infrastructural bottlenecks.

6.1 Introduction

In this chapter, the aim is to put an emphasis on the historical, cultural and socio-economic impact of the transformation process of traditional urban cities into a resilient and sustainable smart city. Section 6.2 explores the information on the transformation process, while Sect. 6.3 focuses on the overview of the related literature on resilient and sustainable smart cities. Section 6.4 analyzes the historical and cultural transformation effort of traditional cities to resilient and sustainable smart urban cities with a case study of Catalkoy-Esentepe municipal region. Readers of this chapter will find useful information on the historical and cultural background of the region, which goes back to BC 3500 of the neolithic age in Vrysi and Kingdom of Vounos. Section 6.5 investigates the economic potential of Catalkoy-Esentepe. This has been noticed during the transformation process, with an impact of increasing tourism investment and a construction boom in the region. Urban mobility has changed the quality of life of local people, in terms of increasing traffic congestion, pollution, insufficient utility services, security and increasing prices. These

V. Yorucu et al., *Modern Approach of Resilient and Sustainable Smart Cities*, Sustainable Development Goals Series, https://doi.org/10.1007/978-3-032-05715-0_6

negative externalities in the region call special attention to prevent such adverse spill-over effects by implementing necessary action plans. In this section, readers may reach valuable information on construction development and new property investment over the last 3 years (2021–2023), total capacity of bed spaces of hotels located in the region, the tourism potential of the region in terms of tourist arrivals, lodgings and employment. The share of sectoral economic activities and employment by each sector have been explained in this section. The chapter concludes with Sect. 6.6.

6.2 Transformation of a City to a Smart City

Among many definitions, one may be able to define an important term within the context of this study, namely urban resilience: a city's capacity to not only sustain its essential socio-economic functions in the event of economic crises and/or social disruptions, but also to have the ability to recover from natural disasters. Unlike coping, where the strategy would be to only handle and manage these stressors, resilience refers more to an adaptive capacity to recover and grow. This term exists to emphasize the ability of urban systems, and its stakeholders (communities, individuals, etc.) to not just survive these challenges, but also to thrive and transform in the face of adversity.

Wamsler et al. (2013) noted that a disaster-resilient city can be understood as a city that has managed to: (a) reduce or avoid current and future hazards; (b) reduce current and future susceptibility to hazards; (c) establish functioning mechanisms and structures for disaster response; and (d) establish functioning mechanisms and structures for disaster recovery" (p. 71).

Being a smart city requires digital transformation. Many cities in Western Europe, Canada, USA and Middle East (Saudi Arabia) have already succeeded such kind of digital transformation. As noted by Mehmet and Yorucu (2024), Singapore, Copenhagen, Seoul, Beijing, Amsterdam, New York, London, Zurich, Sydney, Toronto, Washington DC, Barcelona, Paris, Frankfurt, Berlin, Hong Kong, Dallas, Madrid, Tokyo and Vienna are some of the nations, which are listed as digital cities of the world, which are indexed by Statista (2022). As can be seen, the list is almost entirely of cities from the developed, or First World, clearly reflecting the fact that access to ICT and financial resources is critical. Table 6.1 shown below indicates information about land use and sources of actions taken during the transformation process of smart cities. Among many, public services, economy, governance, citizens responsibilities, mobility, and environment are the most important pillars to be taken into consideration during transformation.

6.3 Overview of Recent Literature on Smart Cities

Bhattacharya et al. (2020), in a normative approach, emphasize that the inhabitants of cities in the developing world must be entitled to an

Table 6.1 Land use and source of actions taken during transformation cities to smart city

Pillars	Items
Services	1. Medical conditions
	2. Accessibility
	3. Security
	4. Educational facilities
	5. Cultural offers
Economy	6. Productivity
	7. Public–private agreement
	8. Business ventures
	9. Tourism
Government	10. E-government
	11. Transparency
People	12. Employment and training
	13. Citizen participation
	14. Citizen integration
	15. Channel of suggestion
Movility	16. Sustainable transportation
	17. Smart traffic control
	18. Pedestrian streets
	19. Smart parking
Environment	20. Environmental protection
	21. Air quality
	22. Paperwork of sustainable resources
	23. Pollution reduction

Source: Peris-Ortiz et al. (2017)

adequate supply of electricity and water, safety and security, clean and pollution free environment, reliable transport facilities, better and affordable medical facilities etc. Building smart cities in a country would require integrated planning social cohesion, such as covering health, education, transportation, demography and biodiversity under social dimension, income, employment within economic dimension, energy, water, land, sanitation and waste management, pollution, all grouped within an appropriate ICT considering not only the environment, but also local culture and heritage. Achievement of sustainable development must embrace economic, social, and environmental gains jointly and simultaneously through the planning system.

Likewise, Haapio (2012) highlighted the importance of social cohesion in the effectiveness of the government agencies to deliver public services and involving citizens in decision-making processes. Considering digitalization, there are several mechanisms which must be undertaken in existing cities to give them intelligence, such as development of ICT infrastructure in the city for data collection, classification, and identification of the rights of various public and private entities in the access of data by different departments, quality management as well as formulation and updating of existing data, which was underlined by Hameed (2019).

In some cities, such as Vienna, a modern electronic system was implemented, controlling the urban growth of the city through updating databases linked to the output of the strategic plan for the city. The city planners also implemented construction management principles when people applied for building permits and licenses. Everyone was obliged to follow the new rules with respect to the development of the city.

As noted in an earlier study by Mehmet and Yorucu (2024), the smart city paradigm has faced many critics. Suartika and Cuthbert (2020) argued that smart cities cannot be considered an independent force in urbanization. Smart cities use technology and innovation to solve urban problems. These are model cities, like the ones in Scandinavia, such as Denmark, whose experience may serve as guidelines for the rest of the urban centers in the world. The principal 'lesson' emerging is the "community-level" demand for environmental standards. i.e., grassroots action for clean air, water, and environment. The most sustainable cities implement municipal action plans to protect the environment to ensure clean air and water, as much as nature. Citizens of urban cities must acknowledge the necessity of environmental protection, demanding sustainable policies at the city level. National action for sustainability typically follows city-level policies. Considering sustainable mobility and transportation policies, Hachette and L'Hostis (2024) emphasized that public authorities work hard to meet targeted objective of sustainable mobility, particularly in large and medium-sized cities. Investing in smart technology has other positive impacts, such as the creation of a safe and resilient city. Sustainability alone is not sufficient, and the concept of resilience is therefore complementary. Smart investments in digital transformation may contribute to the other tasks of the city, such as urban resilience, sustainability, circular, just and ethical cities, as also highlighted by Irajifar and Vu (2024). This is a global task requiring collective action with adequate investment, from the international bodies, such as the EU. Recently, Sharifi and Salehi (2024) pointed out that resilience is a multidimensional and multifaceted concept used for measuring the capacity of a city to respond to shocks and stresses. Mendle and Hartung (2024) accentuated that smart cities are also an attractive concept for private business, private investment, technology providers, and entrepreneurs. They conceptually welcome the expected higher efficiency and innovation capacity of markets and private business, value technological solutions, and often support or encourage private investment.

With respect to the oil-rich countries, it is important to sound a word of caution. As the age of fossil fuels draws to a close, the long-term sustainability of these states becomes questionable. It is possible, of course, that with smart financial planning, the long run, even a few centuries into the future, may be secured. No society, however, can live off interest and dividend payments forever, unless something like the Singapore model is created as a Smart City with a world-class financial center. In Singapore commercial enterprises have been improved through the publication of real-time data on the operation of city services. The latest progress in technology has helped to make banking and financial innovation virtually limitless. Beyond becoming a world-leading financial hub, Singapore has managed to use cutting-edge artificial intelligence with digital sensors and digital control systems that have helped to control traffic, to create building management systems, digital electricity, gas, water meter and other utilities without human power interference and given a chance to ease congestion, reduce city pollution and waste of human and non-human economic resources. As stated by Mehmet and Yorucu (2024), Singapore is arguably the most successful Smart City. Widely used worldwide wireless network, smart phones, and latest information management techniques, have all contributed to Singapore's success story. Considering physical infrastructure, Williams and Charles (2024) mentioned that city planners must provide physical infrastructure support to their citizens to make their city run effectively. Additionally, they reminded that considering increasing demand, city planners have turned to using technology to better assist them in the delivery of services to improve the lives of their citizens and stimulate growth of their local and national economies.

6.4 Historical and Cultural Transformation Effort of Traditional Cities to Resilient and Sustainable Smart Urban Cities: The Case of Catalkoy and Esentepe

Figure 6.1 illustrates the map of protection of environmental, historical and national heritages of Northern Cyprus. This section will explore information on Catalkoy and Esentepe administrative region as a case study. The aim here is to elaborate more thoroughly the environmental, historical and cultural transformation process from traditional urban life to a resilient modern urban city. Readers of this chapter will find detailed information on human life in Catalkoy and Esentepe from the neolithic age (B.C 3500) up to date.

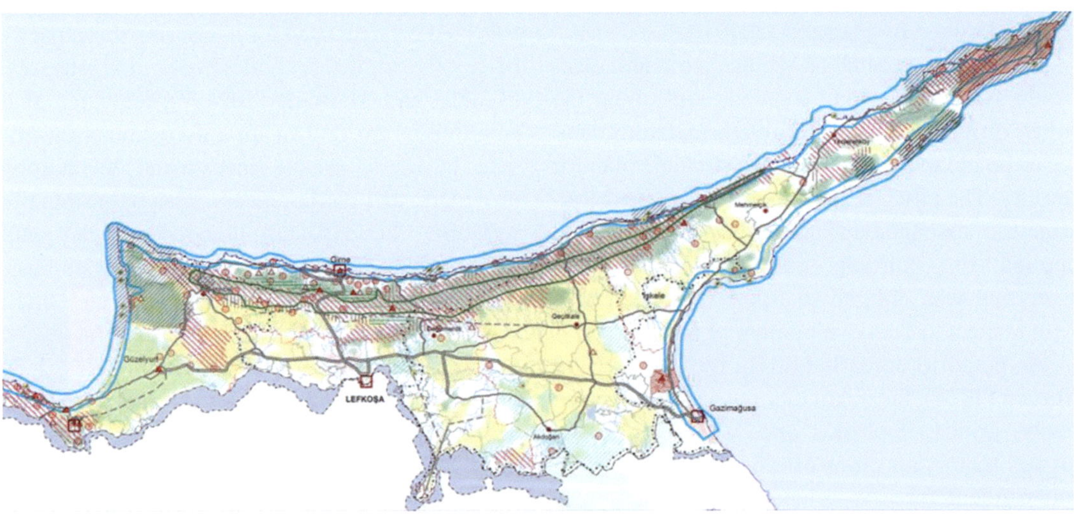

Fig. 6.1 Physical plan of Northern Cyprus portraying the protection of environmental, historical and national heritages of cities located in all districts. (Source: National Physical Development Plan, Urban Planning Department, Nicosia, (2022). Permission granted by Urban Planning Department of Ministry of Interior Affairs, TRNC)

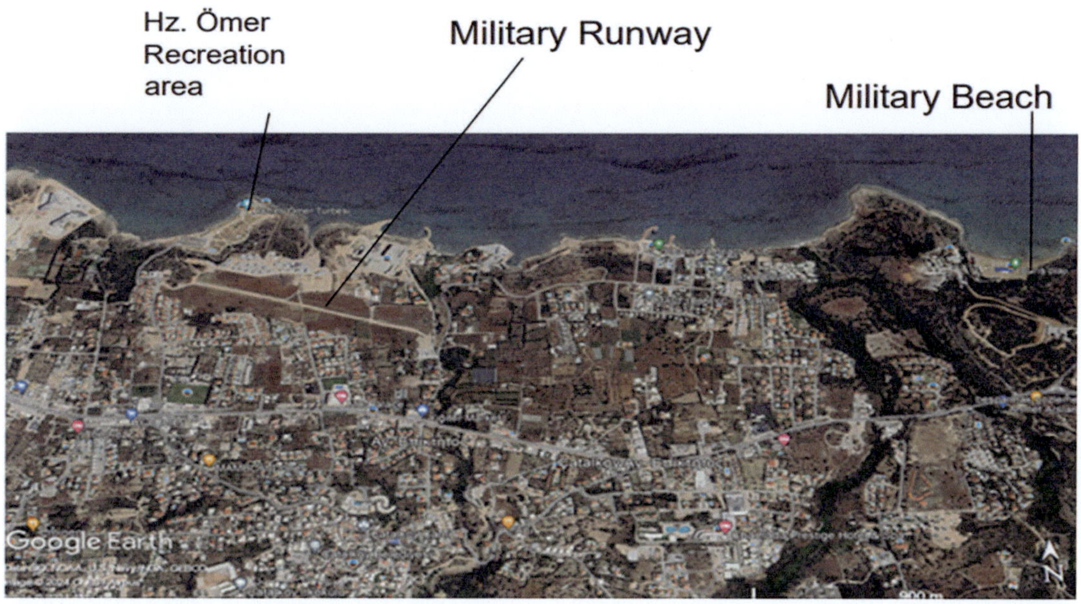

Fig. 6.2 Map of city of Catalkoy. (Source: Google Earth. (2024). https://earth.google.com/web/search/Hazreti+%c3%96mer+T%c3%bcrbesi,+%c4%b0skenderun+Caddesi,+Girne/@35.33597675,33.42224471,1.60062637a,6121.20213671d,35y,0h,0t,0r/data=CiwiJgokCa6vf_uxqkFAEc1lM4EzqkFAGVd37NiotUBAIS1jvHVetEBAQgIIA-ToDCgEwQgIIAEoNCP___________wEQAA. Accessed on 26 Nov, 2024. GoogleAirbusData SIO, NOAA, U.S. Navy, NGA, GEBCO)

6.4.1 Catalkoy Village

Catalkoy (Ayios Epiktitos in old Greek) is a city in Girne (Kyrenia), located at the northern part of Cyprus, and is situated almost 6 km east of Girne (Kyrenia). Figure 6.2 illustrates the Google map of city of Catalkoy, with a location of tomb of Hazreti Omer and his seven companions, old military runway and and military beach (Barıs Plajı) adjacent to each other. The Greek name of the village comes from a German saint who supposedly came to Cyprus in the twelfth century. Ayios Epiktitos (Catalkoy), took its name from a church called the Church of Saint Epiktitos, which was built in 1856 on an earlier site, and contains nothing except an early icon of that saint, who would seem to have been of German origin, a soldier and a companion of St. Auxentios, and to have lived the life of a hermit in a cave beneath the church. Gunnis (1936) also stated that this cave can still be visited by an entrance in the south interior wall which led down a tiny circular staircase to the supposed burial-place of the saint. It contains a ruined tomb and a channel for water. There is also an archaeological site named Vrysi, which is located on a rocky cliff 6.5 miles east of Girne/Kyrenia within the boundaries of Catalkoy (Ayios Epiktitos) village.

Figure 6.3 shows the cadastral map of historical site of Vrysi. Vrysi took its name from the natural spring water, which exists in the area. The historical importance of the site is because of holding first evidence of handmade pottery produced in Cyprus from the last period of Neolithic Age II (3500–3000 BC) and found proof of the first textile making. During the period, there were also remains of grapes, olives and likely that wine making was also known. Additionally, golden earrings, and similar golden artifacts were found on the site indicating that ancient villages had a high level of culture. The findings from the excavation, which was undertaken by British archeologist during 1969–1973 can be seen in the Old Shipwreck Museum of Kyrenia castle.

At a kilometer distance towards the west of Kyrenia, by the seashore is the Tekke of Hazreti Omer (Tomb of Saint Omer), the reputed burial-place of seven Muslim saints ("sahabe"), whose

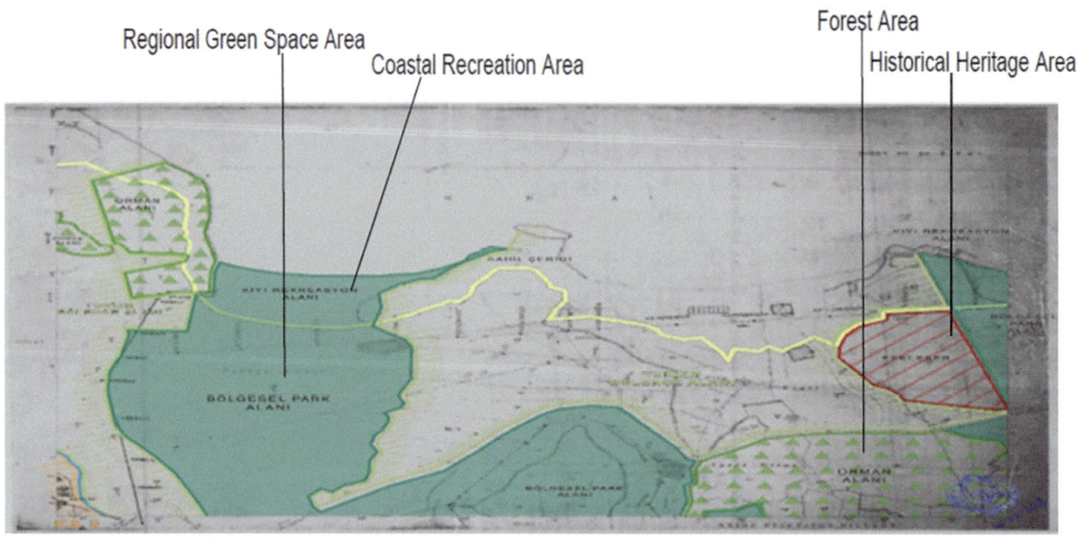

Fig. 6.3 Cadastral Map of Historical Site of Vrysi. (Source: National Physical Development Plan, Urban Planning Department, Nicosia, (2022). Permission granted by Urban Planning Department of Ministry of Interior Affairs, TRNC)

tombs are still shown. This holy burial place represented Moslems interest during one of the four caliphs of Prophet Mohammed, and perhaps the most peaceful location, which is visited by thousands of people every month from all over the world. This area which is shown with Fig. 6.4 has been developed by the municipality of Catalkoy, in conformity with spatial planning. According to the spatial planning of Hazreti Omer, plenty of green spaces have been created. The existing monument has been repaired and maintained with an additional building, a newly constructed modern style mosque for Muslim prayers to practice their religion. The area has also several playgrounds for the children to play in and to do some sport activities. The pedestrians also use the area, which has long pedestrian way with beautiful natural scenery.

The areas depicted by Fig. 6.5 and squared with green dotted lines are allocated for mass tourism investments. Areas with potential ecological, cultural and tourism centers have been spotted with dark red and star signs in the middle, especially for the Vounos area.

There are tiny, ruined chapels, such as St. Stephen, St. Nicholas, and the Prophet Elijah, which are surrounded through the village. On the western part of the village of Catalkoy, there are old king tombs at the location of Vounos Hill. Catalkoy municipality organizes cultural events every year at the Vounos Hill location, which is demonstrated by Fig. 6.6 and shows efforts to make the city a resilient urban city. The Hill of Vounos is a kilometer distance from the Abbey of Lapais, or Bella Pais, which is the most beautifully situated location about half-way down the northern slopes of the Kyrenia hills, some five miles south-east of Kyrenia. Stewart (1908) stated that the Monastery of Bella Pais buildings constitute the finest examples of Gothic architecture not only in Cyprus but anywhere in the Levant. It was built during the reign of Hugues IV. (1324–1339), and was inhabited by French monks, who lived sparingly, but who, despite their jealous affection for their heirs, allowed the place to become ruined. There is another beautiful archeological heritage site on the top of the Five Finger mountains, facing Catalkoy in the North and Arapkoy on the eastern view, called Buffavento Castle. It is situated at a height of 3131 feet above sea level, 14 miles northeast of Nicosia, the capital of Cyprus and was called Lioutas in the past, but the Italians gave it the name of Buffavento (The Defgier of Winds) owing to the many winds blowing there. It is, however, also said that Buffavento is Latinized word for Koutsafentys (the lame

Fig. 6.4 Picture of Hazreti Omer Tomb and Recreation Area. (Source: Author's compilation)

master), the surname of the man who founded the small village in this locality. As emphasized by Keshishian (1980), the village Koutsovendis, the monastery of St. Chrysostomos and the castle are often called after this man. At one time it was also called the Castle of Lion. According to Stewart (1908), the Castle's early history is legendary,

Fig. 6.5 Map of Catalkoy Economic Zone Planning. (Source: National Physical Development Plan, Urban Planning Department, Nicosia, (2022). Permission granted by Urban Planning Department of Ministry of Interior Affairs, TRNC)

Fig. 6.6 Picture of historical and cultural heritage site of Vounos. (Source: Author's compilation)

but its foundation is said to have been due to a lady of noble rank, who, being persecuted by the Knights Templars at the end of the twelfth century, built this fortress on the almost inaccessible summit of Buffavento as a refuge from them. He added that, this lady was also the founder of the Covent of St. Chrysostome.

6.4.2 Esentepe Village

Another village which was united under one municipality is Esentepe/Ayios Amvrosios (in Greek). This village is in the eastern part of both Kyrenia and Catalkoy. It is 29.3 km away from Kyrenia and 22 km from Catalkoy, respectively.

Fig. 6.7 Map of city of Esentepe. (Source: Google Earth (2024). https://earth.google.com/web/search/Esentepe/@35.34338695,33.58671947,154.05694092a,4583.365823 3d,35y,0h,0t,0r/data=CiwiJgokCbWE9fM9q0FAEThLK W a _ q k F A G T Z u 6 _ YhskBAIX8mUlzXsEBAQgIIAToDCgEwQgIIAEo NCP___________wEQAA. Accessed on 26 Nov, 2024. AirbusData SIO, NOAA, U.S. Navy, NGA, GEBCO)

The Google map of city of Esentepe is shown below (see Fig. 6.7). Both villages lay alongside the northern coast of Cyprus and extends up to 200 m high from the sea level, located on the edge of Five Finger Mountain (Pentadactylos in old Greek and Beşparmak Dağları in Turkish).

As noted by Gunnis (1936), the principal church of the village is dedicated to Saint Ambrose and is a vast new building with pepper-pot domes, strangely reminiscent of those on the wings of the National Gallery. Gunnis (1936) also stated that the village was built on an ancient site, as in several cased white stone terrace floors of the Byzantine period can still be seen. Three monasteries lie near the village, of which Antiphonitis is one of the most delightful monuments on the whole island. During the early sixteenth century the Casale of Antiphoniti belonged to the Lusignan family. It is at present the property of the Monastery of Kykko. There is another monastery, named The Monastery of Apati, which lies about a mile distant from Antiphonistis. This was built in the twelfth century and mostly ruined. The third monastery, which is Melandryna is situated close to the seashore. The original Church of Virgin was built in the 15 centuries, but it was repaired in 1731, when large flying buttresses were added externally. The area attracts a lot of visitors as well as settlers to come and work in the tourist resorts. Figure 6.8 illustrates the map of coastal zone of Catalkoy-Esentepe which attracts migration (circled in yellow), location of pollution and forest area. The historical and cultural richness

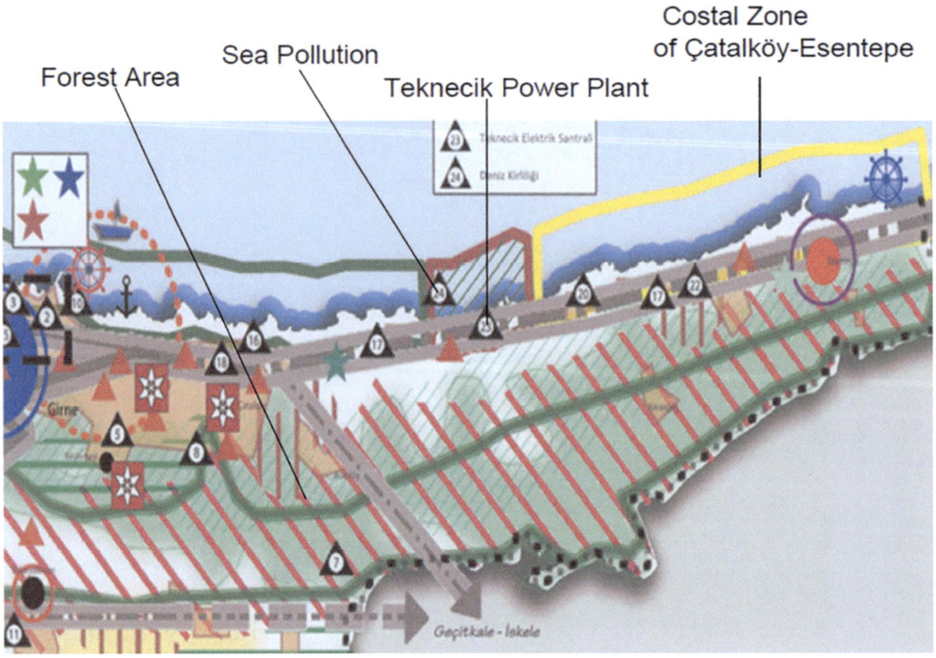

Fig. 6.8 Map of migration area, national heritage sites, areas of pollution, forest and coastal zone of Catalkoy and Esentepe. (Source: National Physical Development Plan, Urban Planning Department, Nicosia, (2022). Permission granted by Urban Planning Department of Ministry of Interior Affairs, TRNC)

of the municipal area of Catalkoy-Esentepe has many tourist attractions which will help to make the city a resilient city, as well as a center of tourism with a lot of economic attractions to generate income to make the city also economically sustainable. Figure 6.9 shows the map spatial planning regions for social entertainment, economic, trade, health and school projects. It needs to be well planned to make it accessible and attract tourists to enjoy daily excursions, while visiting historical heritage sites and bringing economic welfare to the whole region. Figure 6.10 indicates the map of spatial planning for the location of forest area, environmental protection and Alagadi natural life protection zone.

The Cities of Catalkoy and Esentepe possess certain spatial or geographic characteristics (such as sand, sea & sun, good climate, marinas, proximity to a regional market in Turkey and Middle East) conferring comparative advantages that lead to special "niches" in central place location. These advantages lead to economic growth in proportion to the creation of economies of scale in the central place, attracting related industries and services through linkages and interdependencies.

The yellow range of coastlines highlighted (see Figs. 6.8 and 6.11) indicates the tourist development resort of Catalkoy-Esentepe. This area is also open for construction development with certain land zone restrictions, such as land use area and height of the buildings, not more than two floors. Brown rectangular area has been indicated as natural protection zone of Alagadi. The blue circled region indicates marine development zone (see Fig. 6.11), and such area has been planned for investments of marinas. The pink spikes shown on Fig. 6.11 separate zones into areas other than economic investments, such as for green spaces and forests. The grey color lines demonstrate existing highways and intercity highways, as well as a potential expansion plan of intercity transportation. The other details portrayed on the map are out of Catalkoy-Esentepe physical development plan.

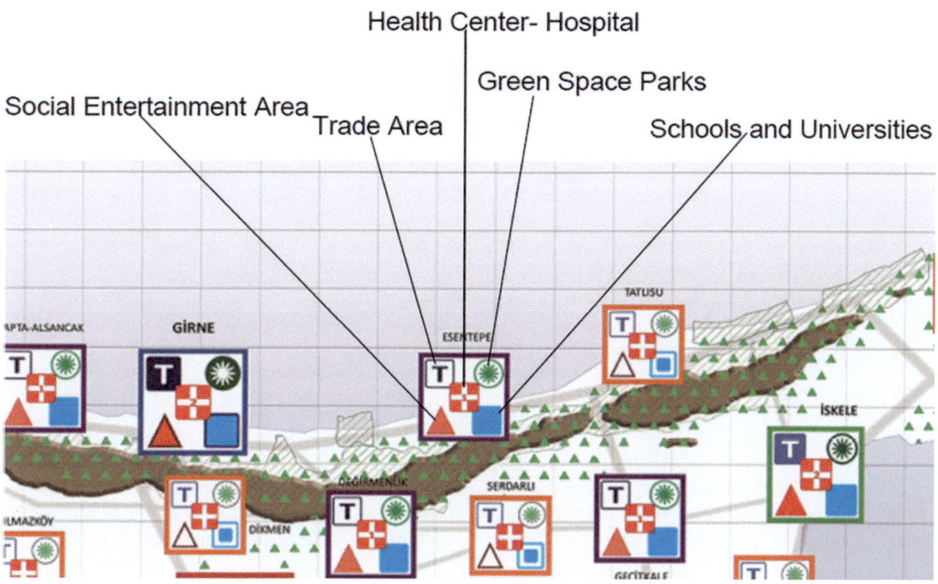

Fig. 6.9 Map of Esentepe Spatial Planning Zone. (Source: National Physical Development Plan, Urban Planning Department, Nicosia, (2022). Permission granted by Urban Planning Department of Ministry of Interior Affairs, TRNC)

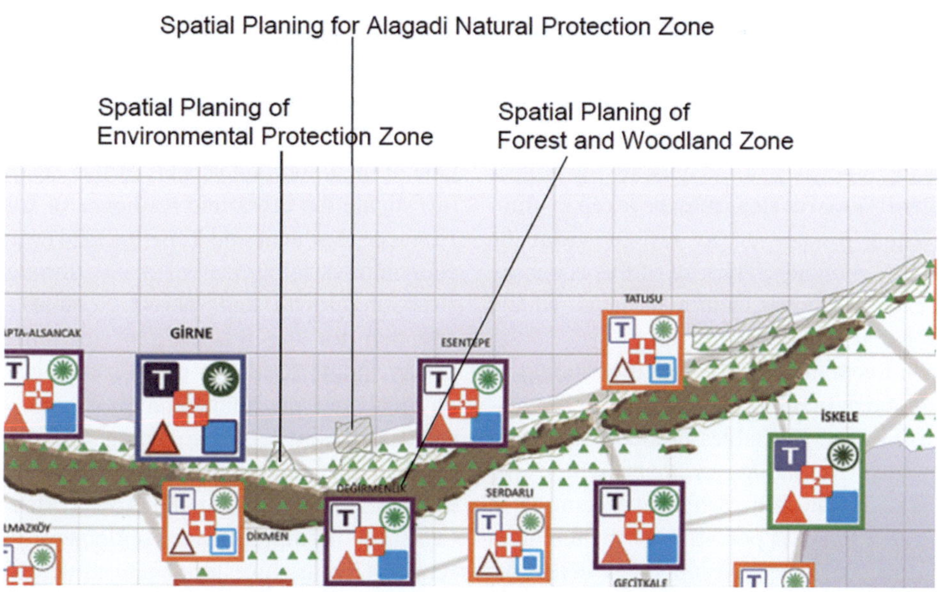

Fig. 6.10 Map of spatial planning zone of Catalkoy: Alagadi Natural Protection Area, Environmental Development and Management Field, Forests and Woodland Park, and Coastal Zones. (Source: National Physical Development Plan, Urban Planning Department, Nicosia, (2022). Permission granted by Urban Planning Department of Ministry of Interior Affairs, TRNC)

Being small is said to be vulnerable, but beautiful. The smallness may be more than compensated for by certain unique characteristics possessed by microstates. Considering the Region of Catalkoy and Esentepe, an additional investment in the region will create labor shortages. International Final University in the region must have labor development programs in their curriculums, organize on the job training programs in tourism and hospitality, gastronomy,

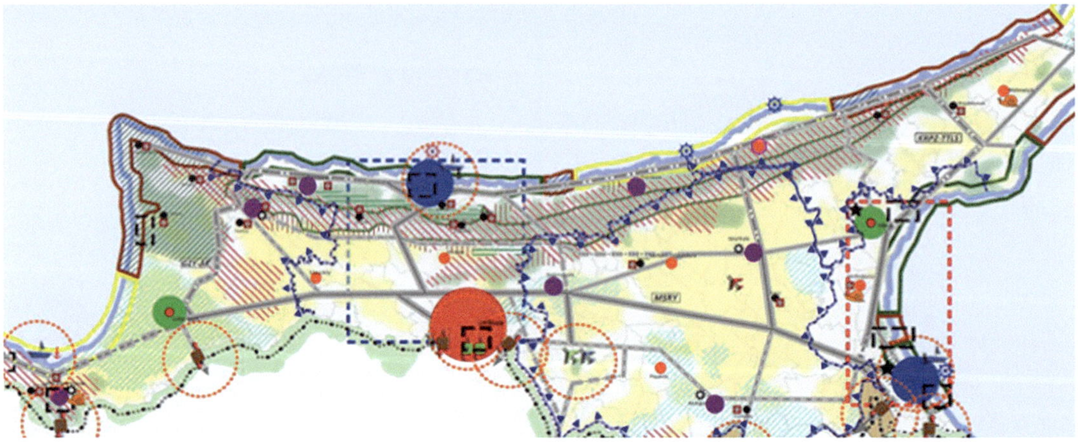

Fig. 6.11 Map of Coastline of Northern Cyprus portraying new construction development zone both in Catalkoy and Esentepe villages. (Source: National Physical Development Plan, Urban Planning Department, Nicosia, (2022). Permission granted by Urban Planning Department of Ministry of Interior Affairs, TRNC)

hospitality, business management, marketing, organizational behavior etc. For the sake of job creation and enhancing labor skills, quality programs should also be developed on ICT, technology use and know-how.

One potential domestic source of financing investment requirements in the public sector is to increase the efficiency of the existing tax collection system. Decisive steps may be taken to eliminate existing leakage of tax revenue from the underground economy. The size of this economy is not exactly known, although it is widely believed that it is quite significant.

Carbon free transportation facilities should be enriched, and agricultural production, in particular fresh fruit and vegetables production, should be supported by the government within the region. The Tarsan farm, which is operating in Arapkoy (Klepini) has potential for the region in terms of food and vegetable supply, as well as for employment. Greenhouse cultivation should be incentivized within the regions of Esentepe and Catalkoy. The Uzundere dam in Arapkoy should be used for agricultural purposes, and development of agrotourism may require additional water. Considering the potential of the region, the amount of water reserved in the dam may be used for irrigation purposes, especially to cultivate high value-added types of agricultural products and fresh vegetables.

6.5 Economic Potential of Catalkoy-Esentepe

6.5.1 Construction Development in the Area of Catalkoy-Esentepe

One of the important aspects of this chapter is to investigate the economic resilience of Catalkoy-Esentepe. Mehmet and Yorucu (2024) stated that economic development of a city through construction growth and tourism requires special attention for the residence of the urban city. Urban sprawl results in many social, economic, physical and environmental problems. A construction boom has been going on since early 2000 in the region of Catalkoy-Esentepe. First boom came out during the Annan Plan period (The UN comprehensive settlement plan) to reunite Cyprus. Investors believed that if Annan Plan would have been accepted, they would hold the ownership (formerly Greek owned property would have become Turkish owned property after being compensated) and the value of assets were expected to be double or even triple. Annan plan was accepted by Turkish Cypriots with 65% yes votes and rejected by Greek Cypriots with 76% no votes. After Annan Plan, North Cyprus has expanded the inter village highways infrastructure and added additional diesel engine turbine transformer units to the existing stock of Teknecik Power Plant. Soon after the expansion of electricity and road networks, Turkey has succeeded to bring 75 million cubic meters of drinkable water from Anamur-Turkey. A subsea water pipeline runs 82 km from Anamur to Gecitköy direction and flows into Gecitkoy dam. Subsequently, Turkey has helped to expand the existing intercity and intracity water distribution system all over Northern Cyprus. Soon after the completion of all physical capital infrastructure, the private sector investments boomed in the region. The Cratos Hotel, Elexsus Hotel, Mc Palace Hotel were built, and Acapulco Hotel, Olive Tree Hotel and Chamada Hotel (formerly known as Malpas Hotel) have expanded their existing capacities. All of them are in Catalkoy. The Korineum Golf Resort was built in Karaagac village. Currently two brand new hotels are under construction in Catalkoy, near Hazreti Omer and the third one is also under construction, which is adjacent to Catalkoy administrative territory. So far, no hotels have been constructed in Esentepe and Bahceli region. Nevertheless, plenty of tourist villas, holiday complexes, a brand-new marina were constructed in Karaagac-Esentepe-Bahceli route and a new wellness health center (Small sized hospital which will also serve some surgical operations) is under construction at Karaagac region see Fig. 6.9). Small scale industry center is operational in Catalkoy. In 2024, an additional two units of diesel engine electricity turbines were installed in Teknecik. The EU funded Solid Waste Transfer Station has been constructed in 2024 (see Fig. 6.12).

Additionally, there is an urgent need for investing in telecommunication and satellite network, ICT, such as smart cameras, smart led super-efficient city lights, censored traffic lights, digital networking and a big server for electronic data collection and storage etc. Such investments need to be projected to transform the city into a smart city. Extra funding is needed for repair and maintenance of schools, leisure centers, sport centers, and public buildings. External finance must be found for spatial planning (see Figs. 6.9 and 6.10), building green areas, and large amounts of finance to revitalize archaeological and historical heritage sites. Other than that, the private sector keeps spending to develop houses, apartment flats or commercial units. According to official statistics (CTSA 2024) total amount of €11,113,760 construction permits were granted by the Municipality of Catalkoy-Esentepe in 2021, €11,894,800 in 2022 and €15,301,668 in 2023 (see Table 6.2 below), respectively. The figures for 2024 are expected to be much higher than the previous years. Additionally, the cost of repair and maintenance of existing or incomplete dwellings, commercial units, hotels, or hotel-apart add up on top of the construction value added realized in the region. For new buildings, massive opportunities are available for Besparmak, Arapkoy, Karaagac, and Bahceli, such as leasing large empty plots to build new five-star hotel complexes, or business centers for the international or multinational companies. Building a new convention center in Esentepe will attract many foreign companies to carry their meetings to the area. On the way to Catalkoy-Esentepe, the Alagadi Natural Park can be transformed into a kind of leisure and sports area with green parks, pedestrian paths, cycle roads and possibly picnic areas. The park accommodates the first quality sandy beach, which is an excellent habitat for Caretta Carretta Turtles. The park can also be used for entertainment purposes.

Esentepe has experienced an economic boom and is an ideal regional hinterland for all kinds of related services. The region, which includes the villages of Arapkoy, Bahceli and Karaagac are suitable for building boutique hotels, restaurants and cafes. On the eastern side, the region expands to Alagadi and Arapkoy. It is an important junction on the expressway to Kyrenia, which is the most important tourist city of the Island.

6.5.2 Tourism Development in Catalkoy-Esentepe

Catalkoy-Esentepe region is one of the most prosperous tourist resorts in North Cyprus. There have been plenty of tourism investments undertaken during the last three decades in the region, such as the construction of five-star hotels,

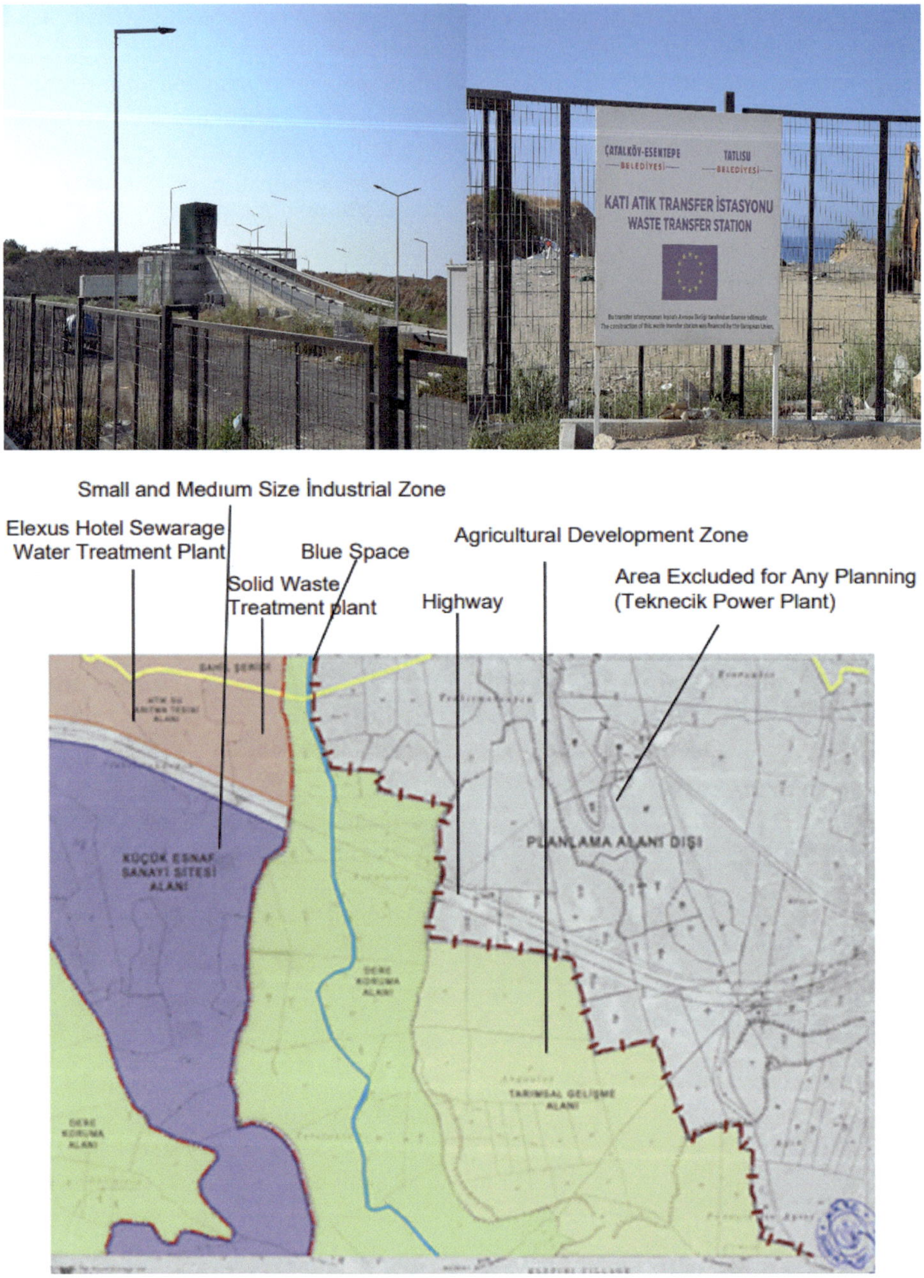

Fig. 6.12 The Cadastral Map of Solid Waste Transfer Plant, Water Treatment Plant¸ Ravine Protection Zone and Teknecik Power Plant. (Source: National Physical Development Plan, Urban Planning Department, Nicosia, (2022). Permission granted by Urban Planning Department of Ministry of Interior Affairs, TRNC. This figure is a magnified version of Fig. 6.11, where Teknecik Power Plant, Sewerage Plant and Solid Waste Transfer station are located. Photos are the author's compilation)

restaurants, cafes and bars, foreign exchange shops, rent-a-car offices etc. With tourism investments, the retail sector has also expanded enormously. In the region, there are seven hotels and three holiday villages (bungalow units) available, and all are operational with 5902 bed space capacities. Table 6.3 tabulates the total number of bed space capacity of hotels and other leisure centers available in the region. There are three brand new hotels under construction and expected

Table 6.2 Construction permits granted by the Municipality of Catalkoy-Esentepe

		2021	2022	2023
Overall total	A	160	137	174
	B	42.167	31.110	22.906
	C	171.707.604	231.353.860	500.211.533
Dwelling units	A	69	75	84
	B	20.694	15.673	20.603
	C	98.497.974	157.987.661	413.866.239
	D	106	122	179
Houses	A	69	63	71
	B	20.694	11.677	13.471
	C	98.497.974	123.897.095	284.831.222
	D	106	64	79
Apartments	A	–	12	13
	B	–	3.996	7.132
	C	–	34.090.566	129.035.017
	D	–	58	100
Commercial units	A	5	1	1
	B	1.211	177	168
	C	4.147.917	1.458.623	3.411.365
	E	5	2	1
Individual shops	A	5	1	1
	B	1.211	48	48
	C	4.147.917	358.092	358.092
	E	5	1	1
Shop on top or in the basement	A	–	–	3
	B	–	129	609
	C	–	1.100.531	11.026.353
	E	–	1	10
Other	A	86	61	61
	B	20.262	15.260	15.260
	C	69.061.713	71.907.576	71.907.576
A. Number of buildings B. Surface area (m^2) C. Value in TL (₺) D. Number of apartments E. Number of shops Total building permits valued in € (Investment in hotels, university, public buildings and infrastructure are excluded)		Total amount of permits 171.707.604 TL €/15.45 TL (end year) €11,113,760 (equiv.)	Total amount of permits 231.353.860 TL €/19.45 TL (end year) €11,894,800 (equiv.)	Total amount of permits 500.211.533 TL €/32.69 TL (end year) €15,301,668 (equiv.)

Source: CTSA. (2024). Cyprus Turkish Statistics Association. Permission granted by Cyprus Turkish Statistics Association of Prime Ministry, TRNC. https://www.ktimo.org/Yayin

to be operational from 2025 onwards. The Turkish origin New Track Trading company is constructing a five-star hotel with 290 rooms and 584 bed space capacities at the 500 m of western side of Hazreti Omer monument. The hotel is having golden sandy beach facility, formerly known as Diana Beach. The second five-star hotel construction is 500 m from the eastern side of Hazreti Omer monument with 253 rooms and 528 bed space capacities and is a very luxurious hotel investment undertaken by AYT Adali Ltd. The hotel was expected to be completed before the pandemic, and due to lock-down, the construction activity stopped in 2020 and restarted in 2024. The third five-star hotel which is under construction, named Lord's Palace Lagoon is just on the border of the administrative territory of Catalkoy and adjacent to Cratos Hotel, which

Table 6.3 Total number of bed space capacities in Catalkoy-Esentepe Region (2024)

Name of the hotels		No of beds
Korineum Golf Hotel		172
Olive Tree Hotel (Oscar Hotel Ext)		270
Elexsus Hotel		1734
Acapulco Hotel		2200
Chamada Hotel (Formerly Malpas Hotel)		374
Mc Palace Hotel		235
Cratos Hotel		820
Meryems Boutique Hotel		12
Alkan Holiday Village		60
National Defense Holiday Complex		30
Sub Total		5901
New hotels under construction		
New Track Trading (Diana Beach)		584
AYT Adali Ltd. (Concorde Hotel)		528
Chamada Hotel (expansion)		326
Lord Palace Lagoon (Catalkoy Hotel)		680
Overall estimated total beds		8019
No of hotels in Catalkoy-Esentepe		8
No of hotels in Kyrenia District		80
No of hotels nationwide		166
No of beds in Kyrenia District		15,224
No of beds nationwide (five-star)		16,886
No of beds in Catalkoy-Esentepe		5901
No of projected beds in Catalkoy-Esentepe		8019
No of employed in tourism (nationwide)		148,211
No of employed in tourism (Catalkoy-Esentepe)		4970
Number of tourist arrivals (2023)		
Turkish	1,522,931	
Foreign	331,559	
	Total Arrival	1,854,490
Number of tourist arrivals (2022)		
Turkish	1,117,675	
Foreign	297,391	
	Total arrival	1,415,066
Percentage increase		31.1%

Source: Municipality of Catalkoy-Esentepe (2024). Permission granted by the Municipality of Catalkoy-Esentepe

will be sharing the same seafront. It is a big hotel complex with 250 rooms and 680 bed space capacity and expected to be operational by the end of 2025. The fourth hotel construction is undertaken by another Turkish investor to expand the existing bed space capacity from 374 to 700 units. The hotel was operational with a different name, previously known as Malpas Hotel. The new investors have changed the name to Chamada Hotel. Soon after all these hotels are completed, the total number of bed space capacity of Catalkoy-Esentepe will reach more than 8000, which is more than half of Kyrenia districts total bed space capacity (15,224). The total number of five-star hotels which are operational nationwide have 16,886 bed space capacity.

The total number of tourist arrivals in 2022 was 415,066 and has risen to 1,854,490 in 2023, which is 31.1% increase over 1 year, respectively. According to the statistics provided by Municipality of Catalkoy-Esentepe, (2024), there are 83 restaurants, bars and cafes operating within the region. Out of a total of 83 establishments, 32 of them operate in Esentepe and 51 in Catalkoy, respectively.

Comparing the capacity of Catalkoy-Esentepe tourist resorts, the size of the hotels regarding number of bed spaces and the share of employment are economically very significant. Currently, 4970 people employed in tourist establishments in Catalkoy and Esentepe, which has a very big impact to the local economy compared to total tourism employment (148,211) in Northern Cyprus. Considering foreign exchange earnings and available job opportunities, the region has become the locomotive of the economy of the northern coast of the Island. Along with tourism, the property market has also been expanded. The value added of tourism, construction and ownership of dwellings (estate sector) has increased very sharply. The increase in demand for property has increased the value of assets. Inflation in the region has risen over the last few years and people cannot afford to purchase a new house to live in, but rather they are obliged to rent. Apparently, rents have also risen due to excess demand, and local people have started to complain about the rapid growth of tourism in the region. Of course, there are some negative externalities that we should not neglect. Increase in investments has also generated population growth and migration problems in the region. Rapid population growth has negative effects, such as increasing traffic congestion, lack of infrastructure, insufficient utility services, pollution, and safety. Criminal cases in the region have increased over the last few decades. The mortality rate in the region has dramatically risen due to detecting an incremental amount of cancer cases, especially within 20 miles radius of Teknecik power plant.

Table 6.4 shows the statistics for total tourist arrivals and overnight stay during 2022 and 2023. As can be noticed from Table 6.4, the number of total tourist arrivals in North Cyprus was 3,576,228 in 2022 and reached to 3,989,371 in 2023, respectively. The total number of tourist arrivals over the period 2022–2023 has risen by 11.55%. Opening the new terminal building of Ercan Airport has helped to accommodate more flights to North Cyprus, especially from Turkey. The number of Turkish tourist arrivals was 2,187,184 in 2022 and has increased by 16.44% and reached to 2,546,802 in 2023, correspondingly. Considering the other nationals, such as foreign tourists arriving and staying in tourist accommodations have remained almost the same, except the Germans. The total number of overnight stays by German tourists in 2022 was 31,612 and has increased by 61.54% and reached 51,067 in 2023. If we consider the British, Greek Cypriots and Israeli tourists, both the number of arrivals and overnight stays have slightly reduced during 2022–2023, which is very minimal in terms of percentage changes.

Table 6.4 Number of overnight stay and tourist arrivals

	2022	2023
No of Turkish overnight stay	707,284	867,914
No of Turkish tourists	2,187,184	2,546,802
No of German overnight stay	31,612	51,067
No of German tourists	270,341	255,845
No of British overnight stay	40,795	39,501
No of British tourists	242,618	203,650
No of Greek Cypriot overnight stay	24,187	21,841
No of Greek Cypriot tourists	48,303	46,441
No of Israeli overnight stay	30,937	30,557
No of Israeli tourists	122,706	112,361
No of other nationals overnight stay	123,831	166,229
No of other tourists	705,076	824,272
Total no of overnight stay of NC	958,646	1,177,109
Total no of tourists in NC	3,576,228	3,989,371

Source: Cyprus Turkish Statistics Association (2024). Permission granted by Cyprus Turkish Statistics Association of Prime Ministry, TRNC

6.5.3 Industrial Development in Catalkoy-Esentepe

There are 22 registered and economically active manufacturing firms available in the Industrial zone of Catalkoy-Esentepe. Other than that, 49 companies are still busy with constructing their factories or commercial units within the same industrial zone. These are all classified as small industry, such as aluminum work, ferrous metals, electric electronic repair, car repair, car rental, olive oil refinery, manufacture of furniture, manufacture of dairy products, bakeries, food and likewise. The total number of people employed in the industrial sector is only 215, which is not significant compared to the tourism industry. The percentage share of employment distribution (see Fig. 6.13) in hotels and restaurants is 80.16%, while retail is the second biggest with 15.05%. The rest is not significant and noticeably the share of industry is 3.47%, construction 3.3% and agriculture 2.22%, respectively.

6.5.4 The SME's, Wholesale and Retail Activities in Catalkoy-Esentepe

As can be observed from Table 6.5, the sectoral distribution of economic activities reveals that tourism sector and small and medium size enterprises (SME) in wholesale and retail trade in Catalkoy-Esentepe represent the lion share of all economic activities. Among others, 339 SME's and 146 tourist establishments operate in the region. The construction sector and industry follow the others with less economic impact.

Considering the share of sectoral employment, Table 6.6 displays the statistics of tourism industry, which is ranked at the top of the sectoral employment league table, with 4970 registered employees working in hotels, casinos, restaurants, bars and cafes. The percentage

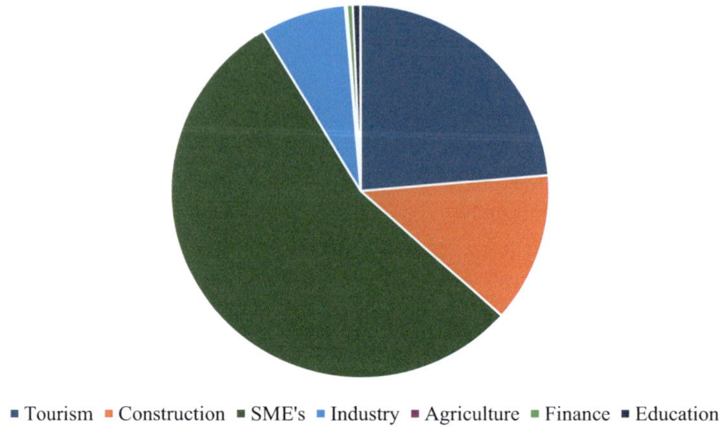

Fig. 6.13 Share of Sectoral Distribution of Economic Activities in Catalkoy-Esentepe. (Source: Municipality of Catalkoy-Esentepe (2024). Copyright permission granted by the Municipality of Catalkoy-Esentepe)

Table 6.5 Sectoral distribution of economic activities of Catalkoy-Esentepe Region

	Catalkoy	Esentepe	Total
Tourism	95	51	146
Construction	50	30	80
SME's	271	68	339
Industry	37	8	45
Agriculture	1	0	1
Finance	2	1	3
Education	3	1	4
	459	159	618

Source: Municipality of Catalkoy-Esentepe (2024). Permission granted by the Municipality of Catalkoy-Esentepe

Table 6.6 Sectoral employment in Catalkoy-Esentepe Region

	Catalkoy	Esentepe	Total
Hotels	2741	175	2916
Casinos	1820	0	1820
Restaurant and bars	128	106	234
Wholesale and retail	780	153	933
Education	260	1	261
Industry	195	20	215
Construction	139	65	204
Agriculture	137	1	138
TOTAL	6200	521	6721

Source: Municipality of Catalkoy-Esentepe (2024). Permission granted by the Municipality of Catalkoy-Esentepe

share of employment in tourism is 73.94%, which is the biggest among all. The SME's available in the region enrolls 933 employees, which is 13.88% of the total employment share of the region. Most of the SME's deal with retail goods and services and a small number are engaged with wholesale construction materials. The share of construction (80 registered companies including developers, contractors, engineers and estate agents) operators and those who are involved in small size industry (45 registered units) are insignificant in terms of employment and economic contribution. There are many construction firms who operate in the region with plenty of labor who are unrecorded, and the central administration is ineffective in controlling the informal economy. The unrecorded economy in the region is massive, as billions of dollars of property investments generated in the region, however the number of registered employees in the construction industry is only limited to the locals, who are only residing within the municipal administration. Foreign workers who work for subcontractors are mainly from Turkey, Africa and Asia, which are not registered with the Department of Labor Affairs of the Ministry of Social Security, and they evade income taxes in an informal sector of Northern Cyprus. The manufacturing sector is at the very infant stage with only 45 registered commercial units, mostly dealing with handy jobs, such as repair and maintenance. Most of the works are based on aluminum metal, PVC and iron works, pluming, woodwork, mosaic and marbles, electric and electronic equipment, bakeries, dairy production and likewise.

6.6 Conclusion

Catalkoy-Esentepe enjoy great potential for "eco-tourism with citta-slow cities" and "bungalow tourism". The region has good sand, sea, and sun facilities. Besides that, it has very good nightlife and entertainment. The region also enjoys more developed physical infrastructure compared to the other municipalities in Mesaria plain and the ones in the western part of the Island. Additionally, Catalkoy-Esentepe has very rich historical and archaeological sites, that is attracting to foreigners to visit and buy properties to reside.

Throughout this chapter it is aimed to focus on how a traditional city can be transformed into a safe, inclusive, resilient and sustainable smart urban city. After reviewing the related literature, Catalkoy-Esentepe municipal region has been chosen as a case-study to investigate the historical, cultural and economic impacts to understand the potential of the region before implementing necessary action plans, policies, strategies and objectives to succeed in the transformation process. The rationale to choose Catalkoy-Esentepe as a case study comparing the other regions in Cyprus is the potential of the region to the developers to build and sell new housing. The foreign demand has increased for the region to buy a summer house from the North, since it is relatively cheaper than the Southern part of the Island. The other reason is the interest of Scandinavians, Iranians, Russians and Kazaks to buy properties from the region relies on the comparative prices of other similar Mediterranean resorts. Comparing the land prices with other similar destinations like Southern France, Italy, Greece or Turkey, north Cyprus is much cheaper. Maldives Homes in Bahceli village are mostly purchased by Norwegians, likewise the Sun Valley homes in Karaagac by the Swedish, and Green Hill in Esentepe by the Bristish and German. Foreign demand has increased the interest of Israeli developers to build and sell. The contractors have already spent millions of dollars to buy and develop land with necessary infrastructure. Such rapid urban sprawl calls immediate attention to making national and regional urban development plans for spatial planning to make the cities more resilient. Both Catalkoy and Esentepe have very rich historical and archaeological heritage sites to promote the region as a sustainable tourist region. After careful investigation, we figured out that the region holds almost 50% of five-star tourist accommodation and lodging facilities and enrolls a significant amount of labor in the service industry. Additionally, there is a large investment incentive program for the development of capacity building in the tourism sector through new hotels and other types of tourism lodging facilities. The incentive program of North Cyprus, which is administered by the State Planning Organization and a certificate of approval provides exemption from income tax for a stipulated number of years, and duty-free imports and other tax privileges to foreign and domestic investors. Other than tourism, the construction sector boomed in the region during the last decade, and the property market became an important source for foreign exchange income as well as employment opportunities to the region. Such economic developments have triggered urban mobility, which increased population growth, the demand for housing and labor force. This has negative externalities to city life, such as traffic congestion, increasing rents, insufficient housing, lack of utility services, pollution, climate change and infrastructure bottlenecks. The mortality rate and detected cancer cases within 20 miles radius of Catalkoy-Esentepe have also increased over the last two decades due to increasing carbon emissions of Teknecik Power plant. The historical, cultural and economic impact of city life has caused rapid urban mobility due to big potential of the city, in terms of economic benefits and employment. This chapter shed light on the basic urgent needs of the city and necessary action plans to be considered to make the city safer and more resilient. Preparations of urban development plans and municipal actions alone are not sufficient. Citizen's participation in planning is also essential to achieve the goal of SDG 11. The implementation of spatial planning, action plans, objectives, strategies, policy recommendations and the assessment needed to succeed the transformation process will be elaborated in detail in

Chap. 7, which will focus on the UN's sustainability agenda and smart city initiatives.

References

Bhattacharya TR, Bhattacharya A, Mclellan B, Tezuka T (2020) Sustainable smart city development framework for developing countries. Urban Res Pract 13(2):180–212

CTSA (2024) KKTC İstatistik Kurumu İstatistik Yıllığı, 2024. Başbakanlık, Lefkoşa, pp 348–349

Cyprus Turkish Statistics Association (2024) KKTC İstatistik Kurumu İstatistik Yıllığı, 2024. Başbakanlık, Lefkoşa, pp 348–349

Google Earth (2024). https://earth.google.com/web/

Gunnis R (1936) Historic cyprus: a guide to its towns & villages & monasteries & castles. In: Rustem K, et al (eds) Lithography in Cyprus, Kyrenia Avenue, Nicosia

Haapio A (2012) Towards sustainable urban communities. Environ Impact Assess Rev 32:165–169. https://doi.org/10.1016/j.eiar.2011.08.002

Hachette M, L'Hostis A (2024) Mobility hubs, an innovative concept for sustainable urban mobility? State of the art and guidelines from European experience. In: Blaid F, Arora A (eds) Smart cities: social and environmental challenges and opportunities for local authorities. Springer International Publisher, Switzerland

Hameed AA (2019) Smart city planning and sustainable development. In: IOP conference series: materials science and engineering, vol 518. IOP Publishing, p 022042

Irajifar L, Vu KN (2024) Digital solutions for resilient cities: a critical assessment of resilience in smart city initiatives in Melbourne, Victoria. In: Sharifi A, Salehi P (eds) Resilient smart cities: theoretical and empirical insights. Switzerland, Springer Nature

Keshishian KK (1980) Romantic cyprus: everybody's guide. Printco Ltd, Nicosia, Cyprus

Mehmet O, Yorucu V (2024) From land disputes to sustainable environmental development: the near east perspective. Springer Publishers, Switzerland

Mendle RS, Hartung A (2024) Wielding a concept with two edges: how to make use of concept with two edges: how to make use of the smart cities concept and understanding its risks from the resilient cities perspective. In: Sharifi A, Salehi P (eds) Resilient smart cities: theoretical and empirical insights. Switzerland, Springer Nature

Peris-Ortiz M, Bennett DR, Yábar DPB (2017) Sustainable smart cities. innovation, technology, and knowledge management. Springer International Publishing Switzerland, Cham, p 72

Sharifi A, Salehi P (2024) Resilient smart cities: theoretical and empirical insights. Springer Nature, Switzerland

Statista (2022). https://statista.com/statistics/1233774/smart-cities-ranking-government-worldwide/#:~:text=Smart%20city%20digital%20capability%20ranking%20worldwide%202022&text=In%202022%2C%20the%20leading%20global,for%20the%20best%20digital%20cities. Accessed 12 Feb 2023

Stewart B (1908) My experiences of cyprus. George Routledge and Sons Limited, London

Suartika GAM, Cuthbert A (2020) The sustainable imperative—smart cities, technology and development. Sustainability 12(21):8892

Wamsler C, Brink E, Rivera C (2013) Planning for climate change in urban areas: from theory to practice. J Clean Prod 50:68–81

Williams DA, Charles CB (2024) 4 IR technologies to facilitate planning in smart cities of the future. In: Blaid F, Arora A (eds) Smart cities: social and environmental challenges and opportunities for local authorities. Springer International Publisher, Switzerland

Internet References

Colonial Office (1893) Cyprus: report on the census of cyprus, taken 6th April 1891, Mediterranean, No. 39. Colonial Office, London. http://www.prio-cyprus-displacement.net

Hart-Davis CH (1922) Report and general abstracts of the census of 1921, taken on the 24th April, 1921. Waterlow & Sons, London. http://www.prio-cyprus-displacement.net

Hart-Davis CH (1932) Report of the Census of 1931. Cyprus Government Printing Office, Nicosia. http://www.prio-cyprus-displacement.net

https://en.wikipedia.org/wiki/Wikipedia:Text_of_Creative_Commons_Attribution-ShareAlike_3.0_Unported_License. Accessed 22 Feb 2023

https://statista.com/statistics/1233774/smart-cities-ranking-government-worldwide/#:~:text=Smart%20city%20digital%20capability%20ranking%20worldwide%202022&text=In%202022%2C%20the%20leading%20global,for%20the%20best%20digital%20cities. Accessed 12 Feb 2023

https://sustainabledevelopment.un.org/post2015/transformingourworld

https://www.siasat.com/saudis-cube-shaped-project-lands-in-controversy-over-replicating-kaaba-2534191/. Accessed 4 Mar 2023

KKTC Başbakanlık Devlet Planlama Örgütü Müsteşarlığı, "15 Aralık 1996 Genel Nüfus Sayımı Sonuçları (Özet), 26, November 1997," Nicosia. http://www.prio-cyprus-displacement.net

Mavrogordato A (1901) Report and general abstracts of the census of 1901, taken on the 1st April, 1901. Government Printing Office, Nicosia. http://www.prio-cyprus-displacement.net

Mavrogordato A (1912) Report and general abstracts of the census of 1911, taken on the 2nd April, 1911. Waterlow & Sons, London. http://www.prio-cyprus-displacement.net

Perry FW (1884) Report on the census of cyprus 1881. Eyre and Spottiswoode, London. http://www.prio-cyprus-displacement.net

Republic of Cyprus (1961) Census of population and agriculture, 1960: Volume I: Population by location, race, and sex, Nicosia

State Statistics Association (2022) İnşaat istatistikleri. Devlet İstatistik Kurumu, Prime Ministry, Nicosia. https://istatistik.gov.ct.tr/TEMEL-%C4%B0STAT%C4%B0ST%C4%B0KLER/%C4%B0N%C5%9EAAT-%C4%B0STAT%C4%B0ST%C4%B0KLER%C4%B0

TRNC 2006 census preliminary results can be found at: www.devplan.org

Wikimapia. http://wikimapia.org/#lang=en&lat=35.000000&lon=33.000000&z=9&m=w

7 UN's Sustainability Agenda and Smart City Initiatives into Selected Cities of North Cyprus

Abstract

This chapter focuses on smart and resilient cities, which is a new paradigm using technology and innovation as a tool of sustainable urban development. This paradigm has been implemented on transforming the municipal city of Catalkoy-Esentepe into an inclusive, safe, resilient, smart and sustainable city. Such a target is not easy to achieve, nonetheless efforts shall be made on the new projects, which are projected by the city planners. In this chapter, city level development plans, action plans, objectives, strategies and policy measures of sustainable urban cities have been examined to reach such a goal. The plans, programs and strategies are subject to be assessed by the key performance indicators to measure the performance of the urban city. Ultimately, readers of this chapter will find information specifically on urban development, housing, green and open spaces, waste management, blue spaces and water management, urban policy and governance.

7.1 Introduction

In this chapter, the aim is to put an emphasis on the implementation of the national and regional development plans of a city during the transformation process. Not only that, but also the action plans, priorities and policy objectives to be followed by the city planners to create an inclusive, safe, resilient, smart and sustainable urban city. Besides that, this chapter explores the legislative and institutional framework of urban development of a smart city. In the next section, we will shed light on the UN Sustainable Development Goal 11. Section 7.3 of this chapter examines smart cities and sustainable urban development. Section 7.4 focuses mostly on the transformation process. In that section, readers may find useful information on city level development plans for the case of Catalkoy-Esentepe. Section 7.5 examines the national development plan, and its governance structure. To succeed in the implementation of national and regional development plans, specific programs and strategies should be made. Each program has its own objectives and policy measures, which are explained in Sect. 7.7. Key performance indicators are needed to assess the rate of success and failure, that is explained in Sect. 7.8. This chapter concludes with Sect. 7.9.

7.2 The UN Sustainable Development Goal 11 (SDG 11)

One of the most important UN sustainable development goals is to make cities and human settlements inclusive, safe, resilient and sustainable (SDG 11). Today's cities and tomorrow's smart

V. Yorucu et al., *Modern Approach of Resilient and Sustainable Smart Cities*, Sustainable Development Goals Series, https://doi.org/10.1007/978-3-032-05715-0_7

cities are intricately linked to the United Nations' 2030 Agenda for Sustainable Development (United Nations 2012, 2015). Sharifi and Salehi (2024) pointed out that in the related smart cities and resilient cities literature some efforts have been made to further clarify the underlying principles of resilient cities and smart cities and define the concept of smart city resilience. According to Sharifi and Salehi (2024), resilience is a multidimensional and multifaceted concept used for measuring the capacity of a city to respond to shocks and stresses. Shcherbina and Gorbenkova (2018) have pointed out that smart city technologies provide efficient city services, as a central instrument of receiving various services, such as energy, water supply, urban infrastructure, transportation etc. for their residents. The related rural development model, which is proposed by Scherbina and Gorbenkova, (op cit) consists of five major systems, such as ecological, economic, administrative, anthropogenic (physical) and social systems. Basiri et al. (2017) underlined that being a place where the virtual and the real meet, merging into augmented urban reality, a smart city is emerging a structured organic whole, single urban entity, a completeness, totality, and integral unity of its parts, components, constituents as technologically, socially, and economically integrated or connected to an urban eco-system. The SDG objective 11 is to make cities and human settlements inclusive, safe, resilient and sustainable, and is a global moral responsibility. Equally, global investment is required to achieve the other SDGs. Irajifar and Vu (2024) emphasized that smart investments in digital transformation of cities at the same time can contribute to the other city agendas as much as possible including urban resilience, sustainability, circular, just and ethical cities. This is a global task requiring collective action with adequate investment, from the international bodies, such as the EU. Mendle and Hartung (2024) highlighted that smart cities are also an attractive concept for private business, private investment, technology providers, and entrepreneurs. They conceptually welcome the expected higher efficiency and innovation capacity of markets and private business, value technological solutions, and often support or encourage private investment.

The flooding of migrants from other cities or other nations can only be controlled by developing a sustainable world. Uncontrolled migration into vulnerable urban cities can only be stopped by providing sustainable development in the world. Fossil fuel-driven automobiles are rapidly ending, replaced by electric cars. Technical knowledge is creating smart cities and super intelligent societies. A greener, and more conservative population is slowly emerging. This chapter has documented market forces as well as city action plans, which can help to facilitate transition to a cleaner, greener global ecology.

7.3 Smart Spatial Planning Principles

As noted by Mehmet and Yorucu (2024), a smart city is a new concept that implements digital information and communication technologies (ICT) to enhance quality and performance of urban life as well as reducing cost. Not only that, but also to control resource consumption, and to engage more effectively and actively with its citizens. Williams and Charles (2024) stated that city planners must provide physical infrastructure support to their citizens to make their city run effectively. Additionally, William and Charles (op cit) addressed that considering an increasing demand in their cities, planners have turned to use of technology to better assist them in the delivery of services to improve the lives of their citizens and stimulate growth of their local and national economies. An early source by Caragliu et al. (2009, 2011a, b) proposed that a city can be defined as smart when investment in human and social capital, modern communication, transportation infrastructure, and promoting sustainable economic development of natural resources through participatory actions and engagements. As also noted by Basiri et al. (2017), smart cities should have characteristics of smart transportation, utilities, healthcare, education, public safety, and smart governance. With the implementation of information technology in smart cities, the waste of natural resources, including energy and water will be minimized, the scarce resources must be allocated efficiently and contribute to reducing carbon emissions. For successful transformation process, the urban city must prioritize the following UN Sustainable Development Goals, such as SDG 7 Affordable and Clean Energy, SDG 8 Sustainable Economy (Sustainable Economic Growth and Employment), SDG 11 Sustainable Cities and Communities, SDG 13 Climate Action (Sustainable Climate) and SDG 16 Peace Justice and Strong Institutions.

Reducing GHG emissions (carbon emission) and achieving energy efficiency is one of the UN's SDG, such as SDG 7 (sustainable energy) and SDG 13 (Climate Change). In addition to that, advancement in technology and utilization of existing infrastructure will improve the quality of life. Besides that, those citizens who travel will benefit from multiple transportation modalities which will be possible through making new services and providing real-time available to guidance to citizens and commuters.

More recently, Zheng et al. (2019) attempted to view Smart City linked to UN's SDGs. For this team, an integrated policy approach is key for smart cities to harmonize the SDG's, especially Goals 11 (Sustainable City and Communities) and 16 (Peace Justice and Strong Institutions) on effective, inclusive, and accountable city institutions and the nexus context. In this integrated framework, different city departments act in coordination across different levels of local authorities to jointly formulate and implement strategies. Thus, the smart city is for resilient citizens, living in quality housing with smart technologies as part and parcel of daily touchpoints that make both huge and small impacts across all walks of people's lives.

The focal point of this section is to focus on legislative and institutional framework for urban development. During the transformation of traditional urban cities to sustainable smart cities, a well detailed national and regional development plan must be prepared. Some may call shortly a master plan of a city or likely they may call "physical development plan of a city". Nonetheless, a city should have certain responsibilities to be fulfilled before the transformation process being completed.

Irajifar and Vu (2024) noted that cities are increasingly using the power of emerging technologies, and digital transformation is revolutionizing different aspects of our societies. Additionally, Irajifar and Vu (2024) stated that the digitalization and the fast transformation of how cities function in the future are inevitable and thus to make building resilience and sustainability keep pace with digital transformations, they need to be integrated for multiple dividend initiatives, by incorporating resilience thinking in smart city projects and digital solutions of building resilience. More detailed information regarding such digital solutions for building resilience is explained in Chap. 9.

At National level, a city must have spatial and urban planning. As also mentioned earlier in Chap. 3, Barcelona's smart city strategy sought to improve city operations, promote economic growth and improve citizens' well-being. This approach aligns with the European Union's Horizon 2020 targets, which may be a very useful example for Catalkoy-Esentepe to learn about their experiences about how they handled spatial planning.

Regional administrations are responsible for developing and implementing spatial and urban planning documents and work in coordination with the national administration. General public services, such as municipal administration, management of municipal properties and issuance of building permits are among some of the responsibilities to be underlined. Public order and safety, public transportation and traffic regulation, road infrastructure and trustworthiness of the roads shall be well planned. Promotion of the city as a tourist resort, identifying its endeavors, planning new recreation and leisure centers are essential during the planning process. Also, the rehabilitation of historical buildings, antiquities and national heritage sites must be well organized. The local economy is also very important and needs to be supported. The Market places should be active and open for regional development. Open marketplaces, fairs, exhibitions, street trade and allocation of natural resources shall be managed. Municipal waste management, including solid waste, paper, plastic, glass, cans, metal and waste of electric electronic equipment (WEEE), street cleaning, public parks, public areas, coastal resorts shall be well directed. Environmental protection is very important and has top priority. Housing and community amenities, such as issuance of building permits, construction permits for non-residential buildings, basic utility infrastructure, local water supply and cemetery services are also other issues which have top priority in urban development. Taking care of food hygiene, public health care, hygiene for sanitary as well as sanitary inspections, and prevention of epidemics are other legislative and institutional items that need to be in order. Local administrations are responsible for all kinds of health-related issues. Elderly care services, including diabetic tests, measuring blood pressures, providing haircut services, assisting disabled people in shopping and travelling etc.

7.4 City Level Development Plans: The Case of Catalkoy-Esentepe

The regional Land Use Master Plan for Çatalkoy-Esentepe municipality was prepared by the Department of Urban Planning in 2022. With such a physical development plan, legal zones and basic parameters for land use have been determined by law in 2022. The Law of Local Administration and Municipalities of Cyprus numbered 51/95 has been changed with a new legislation numbered 15/2022, named Special Law of Uniting Municipalities to reduce number of municipalities from 28 to 18 (Source: Official Gazette (2022) No. 15/2022 Belediyelerin Birleştirilmesine İlişkin (Özel) Yasa, No. 28/2022, 48/2023, 68/2023 and 12/2024 sayılı Değişiklik Yasalarıyla Değiştirilmiş ve Birleştirilmiş Şekli). With the new legislation, which is also called a reform of local administrations, the Catalköy and Esentepe municipalities were united with Arapköy, Besparmak, Karaağaç, and Bahçeli villages. Six villages are positioned under one administration and represented by Çatalköy-Esentepe municipality, which can be observed with the Map of New Territorial Administrative Borders of United Municipalities of North Cyprus (see Fig. 7.1). Such a similar detailed study was undertaken by the UNECE for Tbilisi in 2023 and the UN experts stated that Tbilisi has an ambitious urban development agenda, that is best expressed in the city planning report (UNECE 2023).

Figure 7.2 is the National Physical Plan of Northern Cyprus, which was prepared by the Urban Planning Department of Northern Cyprus. The plan includes all physical infrastructure development needed, including highways, hospitals, schools, green and open spaces, blue spaces, land zoning areas, tourist resorts, agricultural lands, pollution areas, power supply, forest areas, marinas, and etc. The Land Use Masterplan of Catalkoy/Esentepe municipality with an ecosystem-based management approach is consistent with international based practices, which has conformity with scientific consensus statements on marine ecosystem-based management. The Alagadi Natural Resources Protection Plan and the protection of caretta-caretta habitat system is one of the proven types of evidence. Catalkoy-Esentepe land zone master plan provides solution from sustainable forest management to infrastructure-related problems, as well as green and blue spaces infrastructure. Such detailed plan has been prepared by the TRNC Urban Planning Department in 2022, and such National Physical Development Plan indicates Urban Development and land zoning of Catalkoy-Esentepe municipal region, which is highlighted with yellow color (see Fig. 7.2) on the northern coast of the Island, adjacent the tourist city of Kyrenia.

The elected Mayor is responsible for governing six villages as well as implementing the national physical plan prepared by the central administration. Alagadi is a natural park, dedicated as a special protection zone, which is adjacent to Catalkoy. It is an area where the best sandy beaches are available on the Island. The Caretta Caretta turtles spawn on the sandy beach and such habitat needs to be protected. You can see the Google Map (Fig. 7.3) of Alagadi Beach

Fig. 7.1 Map of New Territorial Borders of United Municipalities of North Cyprus. (Source: Urban Planning Department, Nicosia, (2022). Permission granted by Urban Planning Department of Ministry of Interior Affairs, TRNC. https://edevlet.gov.ct.tr/harita-dairesi, https://www.northcyprusuk.com/belediyelerin-haritasi-yayimlandi, https://mevzuat.gov.ct.tr/Portals/48/15-2022%20Belediyelerin%20Birlestirilmesine%20Iliskin%20%28Ozel%29%20Yasa.pdf)

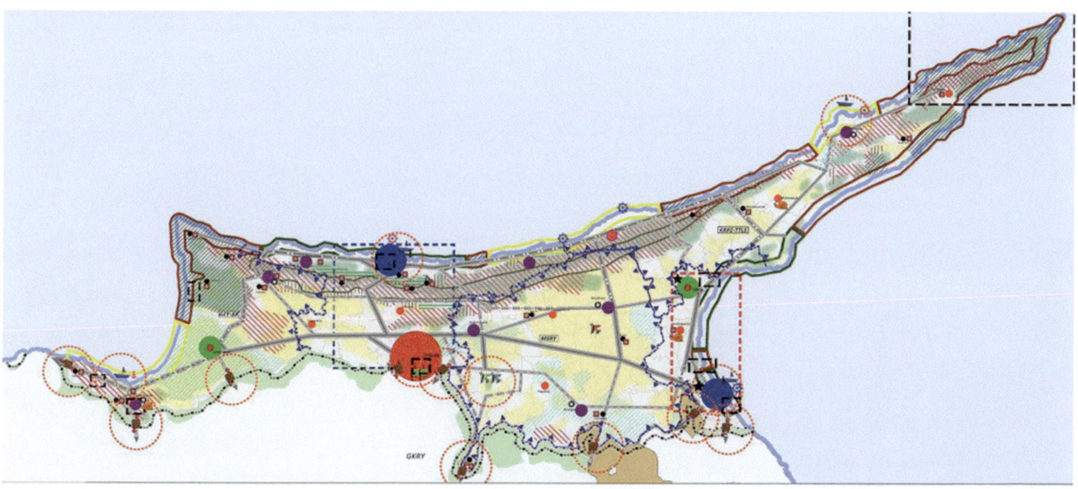

Fig. 7.2 Map of National Physical Development Plan of North Cyprus. (Source: National Physical Development Plan, Urban Planning Department, Nicosia, (2022). Permission granted by Urban Planning Department of Ministry of Interior Affairs, TRNC)

below and a cadastral map of Alagadi (see Fig. 7.4) with its special protection zones that are highlighted with dotted brown lines.

Çatalköy has a recreation area, named Hazreti Ömer, known as the holy place for pilgrimage in Islam. The same region is just a few kilometers away from the main electricity power plant, called Teknecik. Such a plant has two fuel oil turbines and two diesel engines to produce electricity, almost satisfying half of electricity demand

Fig. 7.3 Map of Alagadi Village. (Source: Google Earth. https://earth.google.com/web/search/Alagadi+Girne/@35.33525305,33.48489733,1.65725867a,3660.14307698d,35y,0h,0t,0r/data=CnwaThJICiUweDE0ZGU0MTBkMTk4YWQ3NjE6MHhkY2ZiNWMzYzQ2MDBlYzFmGduMLsrHqkFAIZaLQtzmvkBAKg1BbGFnYWRpIEdpcm5lGAIgASImCiQJmSe-NKytQUARYIE8qzuqQUAZRDFdCJjPQEAh9BX0NpvGQEBCAggBOgMKATBCAggASg0I__________ARAA. 35°20′25″N 33°29′49″E. Accessed on 20 May 2024. AirbusData SIO, NOAA, U.S. Navy, NGA, GEBCO)

Fig. 7.4 Cadastral Map of Alagadi Natural Habitat Protection Zone and Green Space. (Source: National Physical Development Plan, Urban Planning Department, Nicosia, (2022). Permission granted by Urban Planning Department of Ministry of Interior Affairs, TRNC)

annually. Considering carbon emissions, it is one of the most polluted areas in Cyprus, causing environmental risks for the whole island. The rate of GHG emission caused by Teknecik Power Plant (see Fig. 7.5) is the highest in the whole Mediterranean.

Fig. 7.5 Teknecik Power Plant in Catalkoy. (Source: Author's compilation)

Turkey has proposed to lay subsea interconnector electricity gridlines to secure electricity supply and reduce carbon emission to meet the requirements of the EU's 2050 decarbonization path as well as to be in conformity with Kyoto Protocol. Yet, there is very strong opposition by the labor union of KIBTEK, which is the monopoly of state-owned electricity company. Strong opposition and big rallies by the labor union did not help to reduce record high electricity prices and continue to pollute the environment. Since the Northern part of the Island is out of the EU Acquis Communautaire, the European authorities cannot enforce any power to shut the oil turbines. Not only that, but also no one can help to transform such an obsolete electricity turbine into a natural gas turbine or other alternative source of energy.

Black circled triangles depicted by Fig. 7.6 specify the locations of Teknecik in Catalkoy and Aksa in Kalecik. One is situated on the northern coast and the other was situated in the eastern part of the Island, which you can have the best public beaches. Blue striped areas indicate air pollution zones around Alagadi, Karaagac, Esentepe and Bahceli in the north, and bay of Famagusta (Salamis) and Iskele in the east, which have the highest rate of pollution.

Turkey has already completed the feasibility chapter for investing such interconnected grid lines and aiming to start laying the grids in the beginning of 2026. This will also help to produce more electricity from photovoltaic solar parks and by the households on their roofs. The households will then be able to produce their own electricity and sell it back to their excess energy via grid back to KIBTEK and possibly to Turkey. Environmental protection in the region will not only be limited to Teknecik electricity plant, but also Catalköy-Esentepe municipality needs to increase public awareness on solid waste collection and waste management. Plastic, glass, solid food and paper collection containers must be centered all over the place in the city, and the container parks need to be managed by professional experts. Waste of Electric Electronic Equipment shall also be collected and served in licensed depots operating in two different regions of the Northern Cyprus, such as Greenwise Ltd. (The Greenwise Ltd. facility is on the Nicosia—Güzelyurt/Morphou main road) in the west, and Gurdağ Ltd. (Gürdağ Entegre Geri Dönüşüm facility is in Haspolat/Mia Milia industrial area) in the east of Nicosia. The two main WEEE waste operators in Southern Cyprus, who have invested

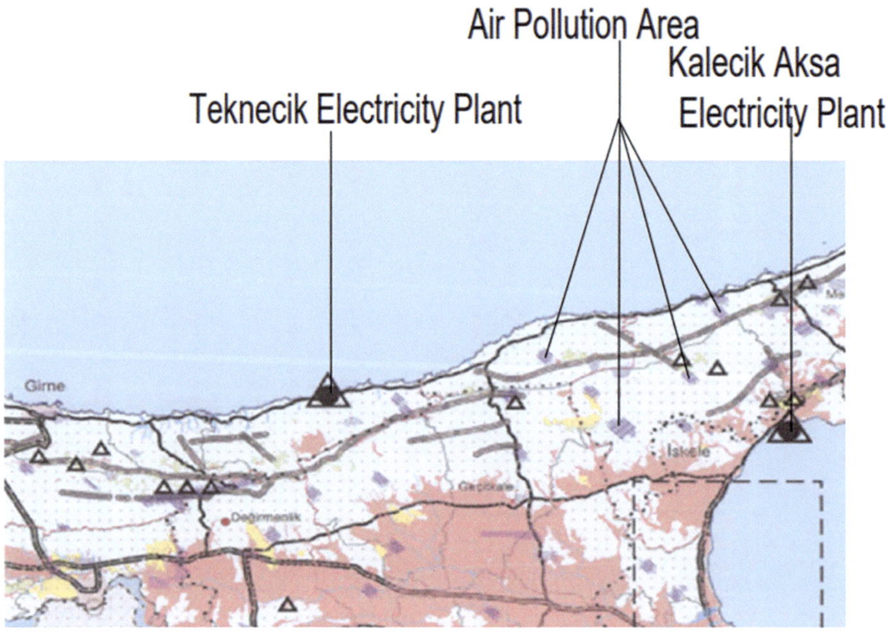

Fig. 7.6 Map of Teknecik Power Plant at Catalkoy and Aksa Power Plant at Kalecik. (Source: National Physical Development Plan, Urban Planning Department, Nicosia, (2022). Permission granted by Urban Planning Department of Ministry of Interior Affairs, TRNC)

in infrastructure for the management of WEEE are Epiphaniou Scrap Metal Ltd. and Economides Metal Recycling Ltd. These two operators have been traditionally the two waste operators that are serving the WEEE Extended Producer Responsibility system in the South (WEEE Cyprus Ltd), for over a decade. They can each treat at their facilities most of the types of WEEE, apart from fridges, which require more specialized and expensive treatment because of the Chlorofluorocarbons in the cooling systems and the insulations of most colling appliances.

The legislation on the national level must be amended to increase the public awareness of recycling and reuse of waste. In the law of Environmental Protection and waste management, there must be specific rules assigning responsibilities both to the Producers' and Consumers'. The waste of the North can then be exported to the South via the Green Line Regulation and such exports will increase the foreign exchange income of the North.

There are many archaeological remains available in the region of Catalkoy-Esentepe. The national heritage protection sites in the region of Catalkoy and Esentepe, such as Buffavento Castle, Vounos, Ayios Epiphanos, Vrysi, Hazreti Omer, Saint Antiphonitis Monastery, Monastery of Apati, Monastery of Melandryna, and several other chapels spread all over the region.

Economic, transport and infrastructural development necessitates the rules and regulations, and directives for the whole region. Catalkoy and Esentepe Municipality must have a resilient and vibrant city strategy, which will provide protected and safe living conditions, healthy natural environment and let everyone plan, and be ready to respond to any challenge. To achieve such a target, an urban development advisory board should be established under the municipal administration. For the sake of sustainability, a participatory budget shall be prepared with an open data strategy.

7.5 National and Local Regulatory Frameworks

In this section, we will focus more on regional development plans for Catalkoy-Esentepe. The Ministry of Interior Affairs is responsible for preparing and implementing such plans at the national level. District Offices in each province act as local administration to implement such plans. Under Provincial district offices, there are municipalities which are responsible for implementing regional master plans, such as for the land development and zoning, and development and the protection of coastal regions, national heritages, antiquities and natural resources. Local administrations are also responsible for implementing rules and regulations to protect flora and fauna, as well as the natural habitat. At national level, the plans aim to achieve inclusive sustainable economic growth, as well as succeeding sustainability of social life and the environment. The legislation of Northern Cyprus is mostly harmonized with the EU Acquis Communautaire, which provide a new impetus for scaling up national legislative and institutional reforms across all sectors and policy areas. Regional administrations shall prepare spatial planning for their municipal cities. Regional development plans and infrastructure must conform with the National Physical Plan, nonetheless special regulation zones which are supposed to be defined in the Development Regulation Plan are expected to be prepared by the municipalities. There are three basic problems to consider while implementing such plans, for instance climate change, population growth and rapid urbanization. Local administrations must have proactive approaches to resolve such potential problems. Addressing such current and future challenges, municipalities should create more balanced and sustainable spatial planning and urban development. Local administrations should provide an online land registry cadastral system with blockchain technology. Mendle and Hartung (2024) noted that blockchain technology is arguably still in its infancy as far as real-world examples are concerned, and even more so when it comes to illustrating its use to improve the resilience of cities. Nevertheless, there is an urgent need to transform all printed cadastral information into electronic cadastral systems. With such a system, it will be very easy to reach electronic cadastral survey, and very simple to produce an interactive map and possible 360° street images.

City climate action plan needs to be prepared by the municipalities. Each municipality must have ex-ante and ex-post climate change impact assessments and coordinate the preparation and implementation of climate mitigation and adoption strategies and action plans. Municipalities alone may not be sufficient to achieve climate change objectives; therefore, close coordination must be established with the Ministry of Environment at national level. Among many objectives, the most important one is to reduce greenhouse gas emissions to at least 35% by 2030 and zero emission by 2050, which is consistent with the EU decarbonization path 2050 climate change action. Considering Catalkoy-Esentepe region in Cyprus, Teknecik Power Plant in Catalkoy causes enormous amount of carbon emissions and needs to be closed immediately. Figure 7.5 shows the air pollution (CO_2 emission) from the turbine of the chimney of Teknecik Power Plant. There is an urgent need to cease such kind of environmental disaster in tourist region of Cyprus. Municipal administration should install thermal cameras with air quality monitoring devices. Air monitoring devices shall be installed all over the region of Catalkoy-Esentepe province. There must be a system for collecting and disseminating environmental monitoring data. Investing to establish an online air quality monitoring system to track progress in reducing air pollution is inevitable. Household investments in photovoltaic renewable energy production are above 38% in the region. The quotas have been full, and the Board of Cyprus Turkish Electricity Association does not grant any permit to install additional PV panels on the roofs due to not having any energy storage facility in Cyprus. Off-grid system is not challenging since storage batteries for households' electricity production is very expensive and there is no recycle for depreciated batteries in Northern Cyprus. To ensure the quality, safety and energy efficiency and climate resilience of urban infrastructure and buildings, there is a need to reform the national system of technical regulations standardization. Policymakers must have conformity assessment rules and existing laws shall be amended and harmonized with the EU Acquis standards. Such kind of move will ensure consumer safety and environmental conservation across all sectors, including construction, real estate, transportation and utility providing services. Each municipality must establish conformity assessment bodies as well as inspection bodies, such as bodies for testing laboratories, service/product certification and inspections, accordingly. Market surveillance of industrial and construction materials on the market falls under the responsibility of the Department of Trade of the Ministry of Economics. Building permits and market surveillance of buildings and structures are the responsibility of national and local administration. Metrology and standardization are also the responsibilities of the same authorities. The standards are determined by law and the adoption and registration of standards in accordance with the TRNC Laws are inspected by the municipalities.

The health care, labor, social security and the management of immigrants fall under the responsibility of national government. Nonetheless, the municipal administrations provide additional health care services, such as nursing for elderly people and they provide health services, such as measuring their blood test, checking diabetes tests, etc. Besides that, they provide haircut services by the barbers and hairdressers etc. Also, the municipality authority provides them with excursion services at least semi-annually.

7.6 Action Plans, Objectives and Policy Measures of Sustainable Urban Cities: The Case of Catalkoy-Esentepe Region and Its Local Plans, Programs and Strategies

Main Objectives of the Municipality of Catalkoy-Esentepe

A. Strengthening civil society participation and engagement
B. Creating a learning organization within the municipality
C. Long-term planning
D. Restricting extensive territorial growth and directing the developments towards the developed inner areas

7.7 The Plans, Programs and Strategies

7.7.1 Creating Green City Action Plan

Objectives:

(a) Promote the use of public transport
(b) Urban Mobility

Policy Measures:

- Low-zero emission business
- Bus networking restructuring
- Building bicycle roads
- Developing small smart traffic systems with digital traffic lights
- Regulation of taxis and (nine-seaters Mercedes Vito tourist vehicles) carrying passengers
- Development of sustainable urban mobility plan
- Rental bicycles and scooters. Bicycle share project
- Developing Hazreti Omer Recreation area
- Creating forestry leisure and sport center
- Developing an Acapulco Woodland Park
- Supporting Besparmak village with Cittaslow tourism
- Protecting Alagadi national heritage/habitat zone
- Developing a parking management system
- Development of pedestrian areas
- Considering the needs of people with disabilities
- Engaging foreign nationals in common leisure areas
- Providing transport connections between city centers and hotels as well as university
- Coordination between transport structures and heritage areas (Hazreti Omer, Vounos, Vrysi, Antiphonides, St Epiktitos, Ayios Amvrosios, Panagia)
- Developing a system of main pedestrian areas

7.7.2 Land Use and Mobility Master Plan

Objectives:

1. Easy and fast access to all amenities of the cities for each resident
2. National and Regional master plan specific objectives
3. Revitalization projects of national heritage sites in Catalkoy and Esentepe region
   (a) Catalkoy: Vrysi Excavation area, Ayios Epiktitos Church (German origin cave of a soldier and a companion of St Auxentios: the life of hermit in a cave beneath the church. It contains ruined tomb and a channel for water), Tekke of Hazreti Omer (reputed burial place of seven Muslim saints, a monument and a mosque), Tiny ruined chapels surround Catalkoy village, such as St. Stephen, St. Nicholas and Prophet Elijah. The tombs of King Vounos. Buffavento castle on top of Five Finger Mountains, which is above sea level 3130 ft.
   (b) Esentepe: Saint Ambrose Church, Antiphonitis, which is the most delighted monastery, The Monastery of Apati, which lies about a mile distant from Antiphonitis and Melandryna Monastery situated on the seashore of Esentepe.
   (c) Bahceli: Armenian Monastery of Kalogreas.
   (d) Arapkoy: Klepini or Dandini (1596), so called Maronite's village (Lebanese village). Byzantine Church of B.V.M.
4. Girne-Catalkoy Land Use Development Plan which was prepared in 2022 and for Esentepe region a regional development plan will be prepared, to the areas of Arapkoy, Alagadi, and Karaagac regions. Another planned work for Bahceli is already continuing. Natural protection zone for Caretta Turtles in Alagadi, Golf Resort expansion project, and a Wellness Health Center in Karaagac are other proceeding projects.
5. Cities anthropogenic and natural environment improvement of the quality of environment, such as waste management, Teknecik power plant and cities recreational system.

Policy Measures:

- Reduce air pollution
- Reduce carbon emission in the construction and transportation sector
- Promote the policy for efficient use of critical resource materials
- Give priority for affordable and reliable housing
- Protection of national heritage sites and archeological monuments
- Improve public transportation
- Consider mixed functional areas to reduce traffic flows
- Promote walking and cycling
- Develop a parking management system
- Develop pedestrian areas
- Consider the needs of people with disabilities
- Engaging foreign nationals in common leisure areas
- Provide connections between city sub-centers: (i) hotels and (ii) university
- Coordination between transport structures and heritage sites (Hazreti Omer and Vounos in Catalkoy and Antiphonities Monastery and Monastery of Panagia in Esentepe)
- Development of a system of main pedestrian areas
- Cleaning ravines for ventilation corridors of the region
- Determining the boundary between the built and unbuilt areas for potential development zones or restricted areas for future construction (cabinet decrees for preventing possible development)
- Avoiding urban sprawling
- Improving natural ventilation by opening forest areas to the public, giving access to the forests and natural green resources around the municipal area
- Avoiding new territories and possibility of creation of new mushroom villages
- Prioritize walking, cycling and using scooters
- Improving citizen access of green areas such as forests in Beşparmak, Bufavento, Arapkoy,

Catalkoy, Esentepe, Karaagac, and Bahceli and provide them sport facilities in such green woodlands, as well as playgrounds for the children

- Creating a connected green network (zero emission buses and tours in the region)
- Improving accessibility of green and blue areas
- Prepare new places for animal care (pets)
- Creating a few marketplaces, leisure/recreation centers and Christmas Market style shopping and entertainment areas where local economic support is provided, and local producers are the main beneficiary ones
- Spatial planning for Karaagac not only to the residents but also to the foreigners living in the region

7.7.3 Resilience Strategy

Objectives:

(a) Develop a city that is thriving
(b) Develop active and inclusive transport systems from/to Kyrenia to Catalkoy/Alagadi Beach and Hazreti Omer/Vounos/Esentepe Golf Resort and Esentepe village
(c) Provide social housing
(d) Provide housing in safe locations to all residents in conformity with Kyrenia/Catalkoy Physical Development plan

Policy Measures:

- Sustainable energy action plan of Catalkoy
- Develop a street manual of cycling, walking and public transport
- Resilient and user-friendly bus networks
- Intelligent transport system—bus stops with online ongoing ICT information
- Establishing a new police station in Catalkoy
- Development of knowledge-based economy of structural engineering and construction in Catalkoy and Esentepe
- Establishing a new fire brigade station in Catalkoy
- Establishing an Early Warning System (as well as radar radio system) in Catalkoy against forest fires, floods, earthquakes, casualties in the sea, and other kinds of natural disasters
- Establishing a City Municipal radio/tv station
- Constructing public schools (i.e. preschool education, education for disabled people (autism, dyslexia, etc.), special care and a school for national vocational qualification etc.
- Constructing a building for elderly care services
- Constructing additional elementary schools
- Building centers for handcrafts, cultural and locally produced products (i.e., CADER)
- Providing better cemetery services
- Constructing an indoor municipal swimming pool and sport center with gym facilities

7.7.4 Housing and Catalkoy-Esentepe Resilience Strategy for 2024–2034

Objectives:

(a) Developing approaches for Catalkoy-Esentepe Climate Change adoption
(b) Protection of healthy natural environment
(c) Improving municipal waste management. The experiences of Tatlisu municipality will lead and good lessons can be learned from their experiences as well as from their, success stories.
(d) Establishing an air quality computer information portal to inform decision-making, which is caused by Teknecik Power Plant and vehicles polluting with carbon emission
(e) Mandatory car inspections for carbon emission
(f) Establishing efficient Traffic system
(g) Installing thermal cameras and Geiger counter devices (GCD) all over the region to measure air quality and possible radiation in the air, as well as other harmful products.

Policy Measures:

- Creating public awareness to close fuel oil turbines in Teknecik Power plant. Helping government to act in coordination and take part in

private-public-partnership to establish more sustainable energy action plan for Catalkoy
- Implement an action plan for public awareness for a sustainable environment:
  - Collection of waste, such as plastic, paper, glass, metal and electric electronic equipment
  - Infrastructure climate change adoption commitment
    Container parks
    Waste management
    Depots
- Establishing a company who is going to be responsible for waste management and monetizing collected wastes through exporting to the Southern part of Cyprus via Green Line Regulation or possibly repairing and selling them as reuse products
- Rehabilitating Catalkoy's and Esentepe's parks for resilience
- Restoring Catalkoy-Esentepe forests including Alagadi natural resource protection zone
- Rehabilitation of Alagadi and Arapkoy water dams (Ayandere and Uzundere) or any kind of other water reservoirs
- Small scale urban parks biodiversity assessment
- Cleaning shores and the sea
- Cleaning up ravines for rainwater to prevent floods
- Fighting against harmful caterpillars
- Improving air quality monitoring and management (against air pollution caused by Teknecik Power Plant)
- Establishing a resilient and user-friendly bus network
- Establishing an intelligent transport management system
- Preparing a pedestrian master plan both in the city and at Alagadi natural park
- Preparing pilot projects for sharing bikes and scooters and developing a 10-km Bike Tour Project at Alagadi natural park

7.7.5 Urban Water

Objectives:

(a) The municipality has urban water management objectives
(b) Catalkoy aquifers and new drilling have provided enough water supply to the region. Exploration work should continue.

Policy Measures:

- Reduce water leakages and replace old and exhausted water distribution pipes
- Revitalization of small ravines and water collection to the dam of Catalkoy and Arapkoy (see Fig. 5.6 Uzundere and Fig. 5.7 Ayandere)
- Reduce water pollution
- Replacing asbestos roofs from old buildings and renew old water distribution pipes
- Improving water distribution system network
- Replacing old water meters with the smart ones
- Persistent checks and inspection of the quality of water extracted from existing water wells
- Preventing water pollution from privately owned swimming pools while cleaning their water

7.7.6 Waste and Green City Action Plan for Catalkoy-Esentepe

Objectives:

(a) Saving water consumption
(b) Improve city resilience to the flood risks and other climate related pressures
(c) Improve recycling of solid waste through technological investments and awareness campaign
(d) Implement comprehensive energy strategy for all buildings and go beyond energy to other sustainable aspects

Policy Measures:

- Establishing water sewage and wastewater treatment plant
- Establishing a sewerage system in the long run
- Repair and maintenance of water distribution system
- Reduction of water consumption per consumer
- Improve logistics of waste collection to reduce costs and relieve road traffic (Such project is already funded by the EU and is under process)
- Plastic, paper, metal, glass and electric electronic equipment waste collection parks
- Container Park collects household products, furniture and fittings, old mattresses, garden appliances, machinery, batteries, oil and petroleum products etc.
- Establishing a container park and providing depot services for waste of electric electronic equipment (WEEE) with hook lift container services and exporting it to the Southern part of Cyprus recycle plants via licensed logistic companies within the framework of Green Line Regulation
- Collection of green waste (grass, stems, organic waste, leaves etc.)
- Establishing solid waste collection points and increasing the number of containers among the regions
- Improving Street Cleaning Services
- Collection of dirty wastewaters from sewage
- Waste management and export to the south Cyprus
- Strengthening regulatory and planning capacity
- Construction of new waste treatment facility (also for green waste) to produce organic fertilizer
- Modernization of waste collection vehicles and all other equipment
- Awareness raising campaign for the public
- Closure and remediation of illegal dumpsites
- Collection of construction rubble and reuse, if possible, in other construction works such as road building or marina construction
- Establishing a solar street lighting network with PV operated streetlights
- Installing PV operated drinkable water dispensers in the main centers
- Installing PV operated waste collection bins
- Installing PV operated refreshing units in leisure centers and bus stops against sun stroke when there is heatwave in Cyprus

7.8 Key Performance Indicators and Importance of Gathering Data for Smart Urban Cities

Assessment of Performance of Municipal Services to become smart sustainable city is subject to data collection from key performance indicators. Every smart city must gather electronic data, disseminate and assess such data to understand how much they succeeded in achieving the required sustainable development goals, as also noted by the UN SDG 17 (Partnerships for the Goals) objectives.

To achieve such goals to construct a smart city, the city shall be instrumented to allow the collection of increasing amounts of data about city life, and data shall be collected in a way that can be amalgamated to make it easy to give an idea about daily life. Table 7.1 demonstrates the Key Performance Indicators for Smart Sustainable Cities. The table tabulates information under three dimensions, such as economics, environment, and society and culture, respectively. Data corresponding to each dimension needs to be gathered, technologically managed, detailed, measurable, and accessible. Technical systems should be established to enable analysis and make immediate decisions whenever actions are needed. The city shall be automated and enable appropriate city functions to be delivered reliably, efficiently, and effectively without direct human interference. Citizens' wellbeing should be sustained, enabling dynamic communities that will spur innovations and growth and improving life quality of society. In addition to that, open and inclusive participation in decision making

Table 7.1 Key performance indicators for smart sustainable cities

Economy dimension	Environment dimension	Society and culture dimension
• Draining/storm water system ICT monitoring	• Air pollution	• Student ICT access
• Traffic monitoring	• Greenhouse gas (GHG) emissions	• Adult literacy
• Open data	• Percentage of households covered by an audited water safety plan	• Electronic health records
• R&D expenditure	• Freshwater consumption	• Physicians
• Patents	• Wastewater treatment	• Health insurance/public health coverage
• Tourism sector employment	• Solid waste treatment	• Cultural infrastructure
• ICT sector employment	• EMF exposure	• Informal settlements
• Wastewater collection	• Noise exposure	• Gini coefficient
• Solid waste collection	• Green area accessibility	• Natural disaster related deaths
• Access to electricity	• Protected natural areas	• Disaster related economic losses
• Public transport network convenience	• Recreational facilities	• Resilience plans
• Transportation mode share	• Renewable energy consumption	• Population living in disaster prone areas
• Shared bicycles	• Residential thermal energy consumption	• Emergency services response time
• Shared vehicles		• Violent crime rate
• Public building sustainability		• Local food production
• Integrated building management systems in public buildings		
• Pedestrian infrastructure		
• Strategic city planning documents—urban development and spatial planning		

Source: UNECE (2023)

processes can be arranged with three partite participation including citizens, policy makers and businesses. They can work together effectively to organize the work needed to be done and manage the life of the city for the benefit of all parties.

The collected data should reflect information on specific topics such as safety, housing, health, food security, education, culture, air quality, energy, environmental quality, public space and nature, waste, water and sanitation, urban planning, transport, public sector, innovation, ICT infrastructure, employment, electricity supply, drainage, buildings and social cohesion. Under each heading, there are several subheadings that needed to be followed. For example, regarding education, the education performance is measured by assessing how many students have good ICT access, the rate of school enrolment, how much percentage of population is holding higher education degrees, and the percentage of adult literacy rate is available in that city. Considering health, they check the number of physicians actively employed, what the percentage of maternal mortality rate is, the life expectancy in that city, the amount of in-patient hospital beds, the percentage of population holding health insurance public health coverage and availability of electronic health records. Concerning energy, they assess the performance of the urban city by checking the amount of per capita KWH electricity consumption, electricity consumed by the public buildings, the share of renewable energy consumption, and the rate of residential thermal energy consumption. Counting waste performances, they assess the amount of solid waste collection, the frequency of collection, and the success of solid waste treatment. Urban cities must have good solid waste management if they want to perform successfully. Another important performance indicator of the urban city is the water sanitation issue. The performance of sanitation is assessed by inspecting the availability of drinking water quality, the amount of freshwater consumption, the condition of physical infrastructure of basic water supply, household sanitation, potable water supply, the availability and the percentage of households using smart water meters, the share of wastewater collection and its treatment, the rate of wastewater loss, water consumption per capita, and the existence of water supply ICT monitoring system. For many municipalities, water supply and solid waste collection are the top priority activities. Besides water and sanitation, drainage is also as important indicator to be taken into consideration to assess city water performance.

It is commonly known that the implementation of control systems toward smart infrastructures increases the need for low-cost and low-energy-consumption underground sensing devices. They also aim to analyze the flood risk reduction potential in the pilot catchment, considering the quality of Narrow-Band-Internet of Things (NB-IoT) cellular network coverage which is a promising wireless communication technology for harsh signal propagation conditions, including underground connectivity.

Another important parameter to be assessed is transportation. The efficiency of city transportation can be assessed by checking the availability of bicycle network, percentage of shared bicycles, scooters and other similar vehicles by the community. Providing efficient public transportation is an essential duty for every municipality. Urban cities must have efficient public transportation networks with alternative modes of transportation. Network convenience is also very important for the passengers. They check the data of transportation mode share and assess the index of travel time travelers used. Of course, especially in big urban cities, traffic is a real problem. A traffic monitoring system is needed to check congestion and time loss during the journey. Therefore, dynamic public transportation information is needed to ease traffic problems in general. Roundabouts slow traffic in many cities. Alternative digital signalization systems are needed to help run traffic more efficiently. Intersection controls instead of roundabouts are proven better for solving problems of traffic congestion. For the need for decarbonization rules, low-carbon emission passenger vehicles shall be encouraged if one needs to have higher performance in an urban city life because it is related with the environmental quality. To prevent environmental pollution, electromagnetic field (EMF) exposure and noise exposure must be controlled. In highly populated urban cities due to high electromagnetic fields there are electrical charges which involve ionizing radiation, that has proven to be harmful and endangering human life. Besides environmental quality, air quality is also important to be measured. Air pollution due to greenhouse gas emissions causes poor life quality. It needs to be well examined, and serious measures should be taken into consideration to prevent air pollution. For public space and nature, urban city performance criteria depend on the accessibility of green, blue, and protected natural areas. Recreational facilities also determine the public space performance of the city. To create such facilities, urban planning is needed. Under urban planning, pedestrian infrastructure shall be well planned. Also, urban development and spatial planning are essential for better urban planning performance. Housing is another performance measuring determinant among key performance indicators, such as informal settlements and housing expenditures are two main parameters to be measured. Safety in urban cities should be well planned. Under safety, the amount of violence crime rate, the number of traffic fatalities, amount of police officers provided by the police service, emergency service availability and response time, fire services, natural disaster related early warning system and number of deaths caused by natural disasters, disaster related economic losses, resilience plans and number of people living in disaster prone areas are important determinants to be assessed in order to determine success performance of an urban city. Buildings, including public buildings, must be compliant with the environment and building materials used for construction must be selectively chosen to make the buildings more sustainable. Public buildings must have installations of building management system (BMS). Such systems use sensitive sensors and devices to monitor and regulate individual systems. Heating, ventilation and air conditioning, lighting, security access are controlled electronically. It is an automated system which increases the performance of using public buildings or other kinds of common areas. To succeed in such kind of performance measures, a city must have good ICT infrastructure. WIFI and fixed broadband subscriptions service should be available publicly to be able to access, gather and disseminate data. Household internet access and wireless large broadband coverage with subscription facilities will help to increase performance of many other indicators mentioned above. Electricity supply should not be interrupted to benefit from such ICT facilities. In some countries, frequent electricity cuts disable proper functioning of ICT infrastructure. Access to competitive prices and reliable electricity supply is very important. Households must be careful to consume electricity during off peak hours; therefore, an electricity demand response system must be available to balance the demand on power grid by encouraging users to shift electricity demand to hours when electricity is more plentiful and gain price advantages in tariffs. Installing an electricity supply ICT monitoring system, and smart electricity meters will help to consume electricity more carefully. Additionally, electricity outage frequency and outage time should be controlled, otherwise there is the possibility of having blackouts. This is needed to know the amount of electricity fed into the electricity grid and amount of electricity consumed. It should be balanced, because the grid system cannot have such a massive amount of storage facility. With renewable energy added to existing grid systems, such as electricity produced by photovoltaic panels and wind turbines, may sometime cause an excess amount of electricity in the grid system and storage problems arise. To increase performance in electricity consumption, low tariffs for off-peak hours may encourage some households to consume electricity when there is over supply in the grid system.

Social Inclusion and Culture are other performance criteria to be measured by the city planners. Cultural infrastructure and cultural expenditure are two parameters to be assessed for the resilience of smart sustainable cities. Considering social cohesion, they also examine the voter participation rate in the local elections, the availability of poverty rate (or as an alternative the prosperity or welfare level of society), the existence of income inequality with Gini Coefficient calculations. They also measure the existence of gender income equity and childcare availability as key performance indicators of the resilience of smart sustainable cities. The UN SDG 5 (Gender Equality and the Empowerment of all Women and Girls) is aiming to improve the gender balance in decision making roles within the municipal administration through the gradual introduction of quotas for key positions is another important governance framework of urban cities.

Locally produced goods should be supported by the local authorities for the sake of the resilience of the city. It is therefore essential to inspect local food production as one of the performance criteria. Locally produced goods also help local employment. In many cities there is a high level of unemployment, which is an indicator of the poor performance of a city. Youth unemployment rate is very important for the future of an urban city. Growing service industry in many cities has proven evidence to reduce natural rate of unemployment. Tourism sector employment, and employment in educational services have helped to increase the resilience and performance of urban cities. There is also ICT sector employment in many smart cities of today that we notice as one of the most important performance indicators. Innovation and technological advances in computer information networks have a positive impact on city life. Issuing patents, continuous financial support for research and development expenditures and government support for the small and medium size enterprises can help to have better performance in smart sustainable cities.

The public sector should play an important role between society and the local authorities. Providing e-government services, such as issuing electronic data for building permits, easing applications for citizen's needs, such as issuing birth certificates, ID cards, passport documents, booking appointments with public offices for other issues, reaching cadastral plans, city plans, road maps, online street maps etc. Reaching open data, collecting and disseminating available data will help to make life much easier. Public sector e-procurement opportunities may also help companies to operate more efficiently.

We should note that some cities have a lack of data availability for some parameters, or limited data to achieve the performance criteria. These are the objectives and goals to achieve better performance on the path to becoming a smart sustainable city.

7.9 Sustainability Protocols and Strategic Roadmap for Implementing Smart, Resilient, and Sustainable Urban Development in Çatalköy-Esentepe

7.9.1 Introduction

This section outlines sustainability protocols and a strategic roadmap for implementing smart, resilient, and sustainable urban development in Çatalköy-Esentepe, Northern Cyprus, based on UN SDG 11, EU best practices, and local governance frameworks. The roadmap integrates green construction, smart city technologies, waste management, energy transition, and participatory governance to align with the National Physical Development Plan (2022) and EU 2050 decarbonization goals.

7.9.2 Sustainability Protocols for Çatalköy-Esentepe

The transformation of Çatalköy-Esentepe into a sustainable, resilient, and smart urban hub requires a holistic and integrated approach, combining policy innovation, technological advancement, and community-driven initiatives. As Northern Cyprus faces pressing challenges—from energy dependence on fossil fuels to urban sprawl and climate vulnerabilities—this framework outlines seven core intervention pillars to guide the region's transition toward EU-aligned sustainability standards and long-term ecological resilience.

1. Legislative and Institutional Framework
   - Harmonize local laws with EU Acquis (e.g., waste management, air quality, renewable energy).
   - Establish a Municipal Resilience Office to oversee climate adaptation, disaster risk reduction, and SDG implementation.
   - Adopt blockchain-based land registries for transparent urban planning.
2. Green Construction and Energy Transition
   - Mandate BREEAM/LEED certification for new buildings.
   - Phase out Teknecik Power Plant (replace with solar microgrids and Turkey's subsea electricity interconnector by 2026).
   - Promote rooftop solar PV with net-metering policies.
3. Smart City and Digital Infrastructure
   - Deploy IoT sensors for real-time air quality, water, and traffic monitoring.
   - Develop a Digital Twin for urban planning (BIM + GIS integration).
   - Launch a municipal app for citizen engagement (reporting waste, energy use, emergencies).

4. Circular Economy and Waste Management
    - Expand container parks for plastic, glass, WEEE, and organic waste.
    - Partner with Southern Cyprus for waste export under Green Line Regulation.
    - Incentivize recycling via tax breaks for households/businesses.
5. Nature-Based Solutions (NBS)
    - Protect Alagadi's Caretta caretta habitat with buffer zones.
    - Restore green corridors (e.g., Acapulco Woodland Park, Beşparmak forests).
    - Implement bioswales for stormwater management.
6. Low-Carbon Mobility
    - Introduce e-bike/scooter sharing and 10-km bike lanes (Alagadi to Esentepe).
    - Convert municipal fleet to electric vehicles (EVs).
    - Establish Low Emission Zones (LEZs) near heritage sites.
7. Cultural and Heritage Resilience
    - Rehabilitate Buffavento Castle, Antiphonitis Monastery as eco-tourism hubs.
    - Digitize heritage sites via 3D scanning and virtual tours.

7.9.3 Roadmap for Implementing Smart, Resilient, and Sustainable Urban Development

The transition of Çatalköy-Esentepe into a model of sustainable urban resilience requires a structured, phased approach that balances immediate actions with long-term systemic change. This roadmap outlines a three-stage implementation strategy (2025–2035) to transform the region through policy innovation, infrastructure modernization, and community-centric development, ensuring alignment with EU sustainability standards and global best practices.

1. Phase 1: Foundation (2025–2027)
    - Policy and Governance
        - Amend municipal bylaws for green construction standards.
        - Establish Resilience Advisory Board (local govt, universities, NGOs).
    - Energy Transition
        - Begin subsea grid connection to Turkey (reduce Teknecik dependence).
        - Install 50+ public EV charging stations.
    - Waste and Water
        - Build 2 container parks (Çatalköy and Esentepe).
        - Pilot smart water meters to reduce leakage.
2. Phase 2: Implementation (2028–2031)
    - Smart Infrastructure
        - Deploy 200+ IoT air quality sensors.
        - Launch Digital Twin pilot for flood risk modeling.
    - Mobility and Green Spaces
        - Complete bike lane network (Alagadi to Hazreti Ömer).
        - Restore Ayandere Dam for rainwater harvesting.
    - Tourism and Heritage
        - Open Antiphonitis Monastery Visitor Center.
        - Certify Alagadi Beach as Blue Flag sustainable site.
3. Phase 3: Scaling (2032–2035)
    - Full Decarbonization
        - Close Teknecik Power Plant (transition to 100% renewables).
        - Mandate net-zero buildings for all new constructions.
    - Circular Economy
        - Achieve 70% waste recycling rate.
        - Establish WEEE repair and resale hub.
    - Resilience Benchmarking
        - Publish annual Resilience Index (aligned with City Resilience Framework).
        - Submit EU Green Capital Award application.

7.10 Policy Recommendations and Conclusion

Catalkoy-Esentepe has been experiencing rapid investment in tourism and a construction boom in housing. Development in the real estate sector has caused rapid urbanization over the last 20 years in the municipal region. Such a boom in investment of land purchase and construction sector has underpinned by rising income levels and marked improvements in urban infrastructure and utility services. Nonetheless, urban development proceeded in the absence of environmental planning and a coherent approach of housing. One of the most important UN sustainable development goals (SDG 11) is to make cities and human settlements inclusive, safe, resilient and sustainable. Rapid urbanization development in Catalkoy-Esentepe has resulted in unplanned urban sprawl, substandard and unaffordable housing, traffic congestion, air and sound pollution, infrastructure bottlenecks, blackouts, unfriendly urban design, lack of green spaces and waste management problem. Over the last 10 years Catalkoy has made important progress in addressing these problems, however lack of finance, migration and population growth in the region, mismanagement by the local administrations and lack of knowledge ended up living in a city without implementation of such urban development plans. In 2022, the TRNC Urban Development Department prepared a national physical development plan, which consists of regional land zone plans to be implemented by the municipal administration of Catalkoy-Esentepe. This chapter also examines the implementation of city action plans, objectives and policy measures to be followed. Moreover, this chapter focuses on key performance indicators that need to be implemented to assess the city's successful performance.

References

Basiri M, Azim AZ, Farrokhi M (2017) Smart city solution for sustainable urban development. Eur J Sustain Dev 6(1):71–71

Caragliu A, Del Bo C, Nijkamp P (2009) Smart cities in Europe. In: 3rd Central European Conference in Regional Science—CERS, pp 45–59. Slovakia. http://www.ekf.tuke.sk/cers/cers2009/PDF/01_03_Nijkamp.pdf. Accessed 18 Apr 2017

Caragliu A, Del Bo C, Nijkamp P (2011a) Smart cities in Europe. J Urban Technol 18(2):65–82

Caragliu A, Del Bo C, Nijkamp P (2011b) Smart cities in Europe, vol 48. VU University of Amsterdam, Amsterdam, pp 1–12

Irajifar L, Vu KN (2024) Digital solutions for resilient cities: a critical assessment of resilience in smart city initiatives in Melbourne, Victoria. In: Sharifi A, Salehi P (eds) Resilient smart cities: theoretical and empirical insights. Switzerland, Springer Nature

Mehmet O, Yorucu V (2024) From land disputes to sustainable environmental development: the near east perspective. Springer Publishers, Switzerland

Mendle RS, Hartung A (2024) Wielding a concept with two edges: how to make use of concept with two edges: how to make use of the smart cities concept and understanding its risks from the resilient cities perspective. In: Sharifi A, Salehi P (eds) Resilient smart cities: theoretical and empirical insights. Switzerland, Springer Nature

Official Gazette (2022) No. 15/2022 Belediyelerin Birleştirilmesine İlişkin (Özel) Yasa, No. 28/2022, 48/2023, 68/2023 and 12/2024 sayılı Değişiklik Yasalarıyla Değiştirilmiş ve Birleştirilmiş Şekli)

Sharifi A, Salehi P (2024) Resilient smart cities: theoretical and empirical insights. Springer Nature, Switzerland

Shcherbina E, Gorbenkova E (2018) Smart city technologies for sustainable rural development. In: IOP conference series: materials science and engineering, vol 365. IOP Publishing, p 022039

UNECE (2023) Smart sustainable cities profile: Tbilisi, Georgia. Report prepared by the United Nations, UNECE, ECE/HBP/220, United Nations Publications. New York, NY

United Nations (2012) World urbanization prospects; the 2011 revision. Department of Economic and Social Affairs, United Nations, New York

United Nations (2015) Transforming our world: the 2030 agenda for sustainable development, New York, NY. https://sustainabledevelopment.un.org/post2015/transformingourworld

Williams DA, Charles CB (2024) 4 IR technologies to facilitate planning in smart cities of the future. In: Blaid F, Arora A (eds) Smart cities: social and environmental challenges and opportunities for local authorities. Springer International Publisher, Switzerland

Zheng L, Kwok WM, Aquaro V, Qi X (2019) Digital government, smart cities and sustainable development. In Proceedings of the 12th international conference on theory and practice of electronic governance, pp 291–301

Internet References

https://edevlet.gov.ct.tr/harita-dairesi

https://www.northcyprusuk.com/belediyelerin-haritasi-yayimlandi

https://mevzuat.gov.ct.tr/Portals/48/15-2022%20Belediyelerin%20Birlestirilmesine%20Iliskin%20%28Ozel%29%20Yasa.pdf

8 Planning, Design and Construction of Smart and Resilient Urban Cities: Development Frameworks, Pathways and Challenges

Abstract

This chapter explores the developmental significance and implications of smart and resilient cities, with a particular focus on the challenges inherent in their planning, design, and construction. It further investigates the potential of artificial intelligence (AI) techniques to improve scenario planning methodologies within the context of smart and resilient urban development. Specifically, it identifies opportunities for AI integration across three critical components of scenario planning: plan generation, scenario generation, and plan evaluation. The chapter also addresses the technical and institutional challenges associated with AI integration and proposes practical solutions, while emphasizing the indispensable role of planning professionals in guiding the AI-enabled decision-making process. In addition, the chapter delineates the objectives and foundational principles of territorial spatial resilience planning and proposes a comprehensive framework that incorporates technical, spatial governance, and operational dimensions. A developmental pathway is articulated across these dimensions, accompanied by strategic recommendations targeting three core domains: (1) strengthening the legal and regulatory frameworks, (2) advancing policy systems, and (3) refining technical standards. These measures aim to support both academic inquiry and practical implementation of land-space resilience strategies. Furthermore, the chapter introduces a structured planning framework supported by ecological pathways and performance indicators to guide the evaluation and advancement of smart and resilient cities. Key policy recommendations include: (1) substantially increasing public financial investments to elevate infrastructure resilience standards; (2) enacting comprehensive emergency management laws and regulations to enhance social governance; (3) prioritizing the deployment of intelligent information technologies to foster innovation; and (4) encouraging multi-level, cross-sectoral collaboration led by public authorities. These strategies collectively aim to promote the development of smart and resilient cities, thereby contributing to high-quality and sustainable urban growth. Finally, the chapter introduces the emerging concept of Smart-Resilient 5.0 from an environmental perspective, outlining its theoretical foundations and identifying priority research areas that characterize the next evolution in smart and resilient urban systems.

V. Yorucu et al., *Modern Approach of Resilient and Sustainable Smart Cities*, Sustainable Development Goals Series, https://doi.org/10.1007/978-3-032-05715-0_8

8.1 Introduction

Cities are complex systems with diverse, interlinked elements and dynamic, nonlinear interactions that create unpredictable behaviors, like self-organization and feedback loops, complicating future projections. Additionally, cities face challenges from extreme climate events, technological disruptions, and rapid population changes, which add further unpredictability. Scenario planning provides a strategic approach for stakeholders to assess how today's decisions may shape future outcomes under various scenarios (Chakraborty and McMillan 2015; Peterson et al. 2003; Hao et al. 2024). This method supports urban resilience by informing planning and management practices that help prevent undesirable future outcomes (Chakraborty and McMillan 2015; Wilkinson et al. 2013; Stojanovic et al. 2014).

The current approach to smart and resilient city planning, design and construction still faces shortcomings in its capacity to endure unforeseen large-scale risks and respond to significant emergencies such as natural disasters (Shen and Zhang 2020). This underscores the urgency of integrating smart and resilient city frameworks to comprehensively boost urban resilience and advance modernization, digitalization, standardization, and sustainability in urban management. Emerging technologies, such as big data and artificial intelligence (AI), form the backbone of smart and resilient cities, enhancing the ability of cities to react swiftly, resist disasters effectively, and recover quickly, allowing them to return to their pre-disaster equilibrium state as soon as possible (Shen and Wang 2018; Chen et al. 2021).

Smart cities, by design, focus on improving urban intelligence during routine operations, while resilient cities emphasize sustainability in response to changes brought on by human and natural factors (Xu et al. 2018). It is generally accepted that resilient cities, functioning as complex coupled systems, require multi-stakeholder collaboration (Zhao et al. 2020). With extreme weather conditions and emergencies becoming more frequent, integrating smart and resilient city models is crucial to strengthening overall urban resilience, improving modern governance, and reinforcing sustainability (Xuesong et al. 2022).

In terms of global practice, lessons can be drawn from developed nations' experiences in building digital resilient cities. These insights have laid the groundwork for offering preliminary recommendations on building resilient cities in developing countries. Despite the extensive research on this topic, there are still gaps in specific implementation pathways for integrating smart and resilient city models, particularly in the context of developing countries. Overall, building smart and resilient cities requires an interdisciplinary approach and global collaboration to create sustainable urban environments capable of withstanding and recovering from unforeseen crises.This chapter discusses the implications and developmental importance of smart and resilient cities while also addressing the challenges encountered in their planning and design construction. Besides it examines how AI techniques could enhance scenario planning practices for smart and resilient cities, identifies potential AI applications within three main components of scenario planning: plan generation, scenario generation, and plan evaluation, addresses integration challenges, proposes potential solutions, and emphasizes the essential role of planning experts in guiding this process. Moreover, it outlines the objectives and overarching principles of territorial spatial resilience planning, and construction and develops a comprehensive framework encompassing the technical, spatial governance, and operational dimensions and summarizes a development pathway across these dimensions and proposes strategic measures in three key areas: enhancing the legal and regulatory framework, policy systems, and technical standards, to guide research and practical implementation in land space resilience planning. Furthermore, it proposes a structured framework, ecological pathways, and assessment indicators for the development of such cities. Recommendations to enhance smart and resilient city construction include: (1) significantly increasing public financial investment to raise infrastructure standards, (2) formulating comprehensive emergency laws and regulations to improve social governance, (3) prioritizing intelligent information technology to drive innovation, and (4) promoting multi-dimensional collaboration and governance led by the government. These measures are intended to foster the development of smart and resilient cities, thereby supporting high-quality, sustainable urban growth. Finally, as the next Level of Smart City, the concept of Smart-Resilient 5.0 from an environmental perspective and prominent research areas in Smart-Resilient 5.0 will be introduced.

8.2 The Concept and Development Significance of Smart and Resilient Cities

8.2.1 The Concept of Smart and Resilient City

A smart city integrates emerging technologies to enhance urban infrastructure and improve governance in areas such as human resources, social economy, and traditional industries, aiming for efficient management of natural resources (Meng and Wu 2022). On the other hand, resilient cities are designed to actively withstand disasters and ensure rapid post-disaster recovery through strategic resource allocation (Ribeiro and Gonçalves 2019). Smart resilient city represents the combination of both concepts, blending the technological advancements of a smart city with the disaster-resistance capabilities of a resilient city. This integration allows for continuous intelligent urban operations under normal circumstances and swift recovery in the aftermath of disasters. In terms of structure, smart cities provide the technological foundation, such as data, hardware, and analytical tools, while resilient cities enhance the capacity for rapid response, governance, and recovery. Together, they support emergency disaster prevention and offer broader application scenarios for urban intelligence. While many existing emergency systems in smart cities primarily focus on post-disaster recovery, smart resilient cities take a more comprehensive approach. They emphasize proactive risk resistance, enhancing early warning systems, emergency decision-making, and the ability to learn and adapt from disasters (Ali et al. 2023; Xuesong et al. 2022).

From a governmental perspective, smart resilient cities improve urban governance, disaster prevention, and post-crisis recovery. Key components include resilient transportation, energy, infrastructure, and public services. At the societal level, these cities enhance services related to people's livelihoods, such as community resilience, healthcare, environmental protection, and water services, ensuring citizens' safety and adaptability during emergencies (Xiong et al. 2022a; Xuesong et al. 2022). Smart and resilient city model is illustrated in Fig. 8.1.

8.2.2 The Significance of Smart and Resilient Cities Development

The development of smart and resilient cities emphasizes the integration of advanced technologies with urban resilience strategies. This approach aims to improve cities' capacity to withstand and recover from natural or human-made disasters while enhancing daily urban management through smart systems. By combining the strengths of smart cities and resilient cities, urban areas can effectively respond to social development needs and improve residents' quality of life, creating a safe, healthy, and sustainable living environment (Varzeshi et al. 2024; Xuesong et al. 2022).

Historically, cities have relied on various stages of technological development to enhance their functionality. Digital city was built around databases and information systems. It utilized information technology to offer urban planning tools and predictions, providing significant advantages in e-government, disaster relief, environmental management, and transportation. However, the smart city brought further advancements by integrating the Internet of Things (IoT) and cloud computing into urban operations. This new model enabled real-time data analysis and better coordination across various sectors, includ-

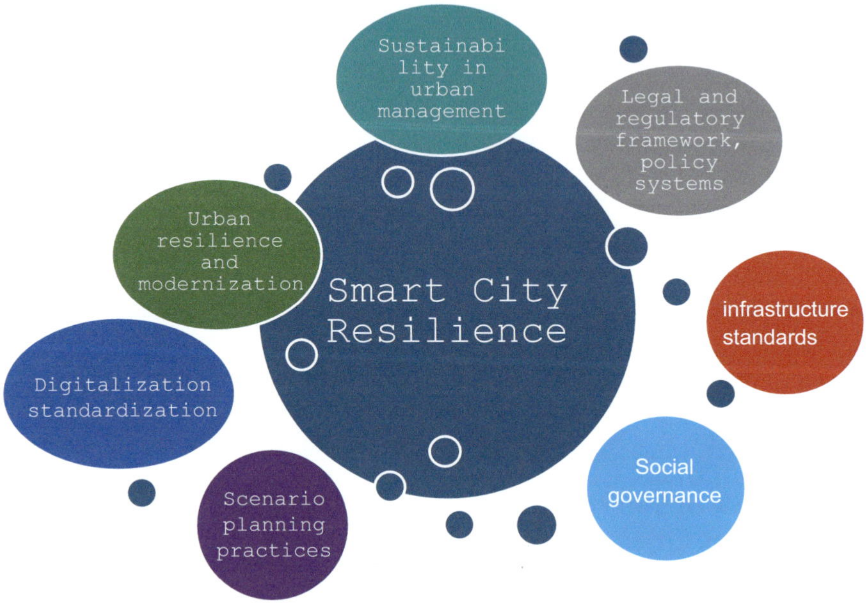

Fig. 8.1 Smart and resilient city model. (Source: Author's compilation)

ing environmental protection, public safety, and urban services. More recently, the new smart city has emerged, focusing on the deep integration of information technology and urban modernization. This model seeks to create efficient governance systems, promote green and open-source economic development, and enhance cybersecurity. By doing so, the new smart city contributes to coordinated development between cities and national policies. Finally, the concept of a smart resilient city builds on the foundation of these earlier models. It incorporates big data and resilient development strategies to enhance cities' ability to predict, prevent, and respond to disasters (Varzeshi et al. 2024; Xuesong et al. 2022).

8.2.3 The Need for Smart Resilient Cities

8.2.3.1 Preventing and Managing Public Emergencies

With increasing urbanization and globalization, the concentration of populations in urban areas has created more complex and interconnected systems that are vulnerable to disasters. These disasters can range from natural events like earthquakes, floods, and extreme weather to manmade crises such as resource shortages, social unrest, and economic disruptions. Addressing these issues requires cities to improve their resilience by enhancing their infrastructure's capacity to absorb shocks and quickly recover (Almulhim 2025; Ribeiro and Gonçalves 2019).

The construction of smart resilient cities supports this need by leveraging emerging technologies such as big data, AI, and IoT to improve urban systems' perception, response, and coordination capabilities. These technologies allow cities to monitor and respond to potential risks in real time, reducing the impact of emergencies and accelerating recovery efforts. Furthermore, they enable better decision-making by providing a comprehensive view of the city's infrastructure and resource distribution during crises (Almulhim 2025).

8.2.3.2 Promoting Economic and Social Digital Transformation

Smart resilient cities also play a crucial role in driving the digital transformation of economic and social governance. As cities grow, the traditional methods of governance often fall short of addressing the complex needs of urban populations. Many cities still face gaps in their administrative, ecological, and public utility management systems, hindering their ability to effectively serve their residents. The digitalization of these systems allows for more efficient resource utilization and coordination, creating a more adaptive and responsive urban environment.

Additionally, the integration of digital technologies in urban governance provides opportunities for cross-regional and cross-level cooperation. This is particularly important in bringing the gap between urban and rural areas, where disparities in access to services and infrastructure can exacerbate social inequalities. By creating a more inclusive and digitally connected urban framework, smart resilient cities contribute to a more equitable distribution of resources and services, fostering social stability and economic growth (Qi and Xiao 2020; Viale Pereira et al. 2023). The economic and social transformation of smart and resilient cities for adaptive responsive resilient urban environment is illustrated in Fig. 8.2.

8.2.3.3 Integrating Smart Technology into the Construction of Resilient Cities

The success of smart resilient cities lies in their ability to seamlessly integrate smart technology into urban resilience strategies. By using technologies such as big data, cloud computing, and AI, cities can gain a comprehensive understanding of their resilience status and make informed decisions on risk prevention and emergency management. This integration allows for the creation of efficient early warning systems, real-time operational adjustments, and the timely allocation of public resources (Ali et al. 2023; Varzeshi et al. (2024).

One of the key aspects of smart resilient city development is the ability to harmonize "soft" (digital and informational) and "hard" (physical infrastructure) elements. In the spatial dimension, this "soft-hard interaction" ensures that cities can effectively respond to emergencies by combining technological solutions with robust physical infrastructure. For example, resilient transportation systems, public services, and

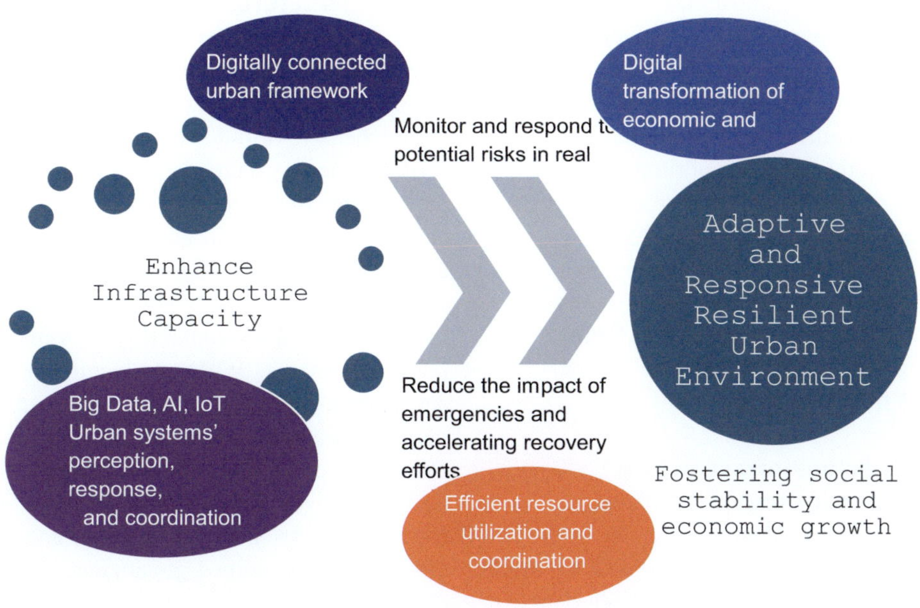

Fig. 8.2 Economic and social transformation of smart and resilient cities for adaptive responsive resilient urban environment. (Source: Author's compilation)

energy networks can all benefit from the real-time data provided by smart technologies, allowing for quicker recovery and continued functionality during and after disasters (Ali et al. 2023; Xuesong et al. 2022).

At the organizational level, smart resilient cities promote refined management during normal times and collaborative governance in times of crisis. This dual approach ensures that cities can maintain efficient operations while also being prepared to handle unexpected events. The use of smart technologies in disaster responses not only improves cities' ability to withstand shocks but also enhances their capacity to learn from past events and adapt to future risks (Almulhim 2025; Varzeshi et al. 2024).

The development of smart and resilient cities represents a critical step in addressing the challenges of modern urbanization. By combining the intelligence of smart cities with the resilience of disaster-prepared cities, this integrated approach offers a comprehensive solution to the growing complexity of urban governance. Smart resilient cities leverage advanced technologies to improve daily urban management, enhance emergency response capabilities, and create a more adaptive, equitable, and sustainable urban environment. The ongoing evolution of smart city technologies, coupled with the increasing need for resilient infrastructure, will continue to drive innovation in this field, supporting the long-term growth and stability of cities worldwide (Almulhim 2025; Varzeshi et al. 2024).

8.2.4 Obstacles to Building Smart and Resilient Urban Environments

8.2.4.1 Deficient Infrastructure Coverage and Insufficient Resilience Awareness

Urban infrastructure is not limited to smart devices like sensors, 5G networks, big data centers, and industrial Internet. It also includes essential public utilities such as healthcare services, road networks, and energy pipelines. These elements are crucial for maintaining the normal operation of urban systems, and their coverage significantly impacts how quickly and accurately a city can sense disasters. In turn, this affects a city's decision-making and its ability to respond effectively during crises. However, many cities have imbalanced infrastructure configurations. Infrastructure tends to be concentrated in central urban areas, leaving remote or less-developed regions with outdated or insufficient systems. Even within central areas, much of the infrastructure is aging and does not meet modern needs for disaster sensing and resistance. Therefore, it is necessary to invest in the construction, maintenance, and enhancement of smart and resilient infrastructure, particularly in small and medium-sized towns. Increased awareness around disaster preparedness, supported by technologies like multidimensional monitoring and intelligent network systems, would greatly improve a city's ability to deliver timely warnings and facilitate efficient coordination during emergencies (Long et al. 2025; Xiong et al. 2022a; Xuesong et al. 2022).

8.2.4.2 Ineffective Governance Structures and Insufficient Resilience in Decision-Making

The governance of urban resilience faces several critical challenges, one of which is the lack of a comprehensive top-down management system. Without clearly defined responsibilities for all parties, the effectiveness of emergency responses can be compromised, leading to uncoordinated efforts in times of crisis. Furthermore, the government often lacks a robust framework for supervision and accountability in implementing relevant systems. Without objective evaluation and assessment mechanisms, urban emergency plans and disaster mitigation decisions lose their effectiveness. This results in weak support functions for governance, limiting a city's capacity for disaster prevention, mitigation, and recovery efforts. A more refined governance system with clear responsibilities and performance evaluations is crucial for enhancing the decision-making capacity during emergencies (Brandt and Siebert 2022; Leitner 2023; Marin et al. 2024; Xuesong et al. 2022).

8.2.4.3 Inadequate Data Integration and Constrained Resilience Response Capacity

While smart city development has progressed rapidly, there remains a significant disconnect between the goals of smart cities and resilient cities. Many smart tools are dispersed and decentralized, making it difficult to integrate vital resources needed for dynamic governance. This disconnection hampers the development of a comprehensive emergency management system, leading to information barriers that restrict collaborative disaster response efforts. Moreover, the integration of smart governance systems established by the government with local community data remains ineffective. Many communities continue to rely on traditional methods of data collection, such as manual surveys, which delay responses and affect the overall efficiency of urban governance. The inability to connect smart systems across different layers of society reduces the timeliness and effectiveness of urban resilience efforts. Achieving seamless integration of information systems would significantly improve the ability to predict, manage, and respond to disasters in real-time (Ali et al. 2023; Xuesong et al. 2022).

8.2.4.4 Inadequate Distribution of Public Resources and Poor Coordination of Resilience Efforts

The allocation of public resources is a fundamental aspect of urban resilience. However, many developing cities still struggle with optimizing emergency management facilities due to limitations in human, financial, and material resources. Communication networks, for instance, often lack robustness, and emergency supply reserves are inadequately managed. When disasters strike, insufficient emergency supplies can lead to severe economic losses and prolong the time it takes for cities to return to normal operations. Inadequate resource allocation also impacts social systems, as seen in the imbalance in public health emergency facilities during the COVID-19 pandemic. This pandemic highlighted the significant disparities between cities in terms of emergency preparedness, with some cities lacking multifunctional public spaces that could serve as flexible, resilient hubs during emergencies. Collaborative governance between various urban stakeholders is another weak point. Responsibilities are often fragmented across multiple agencies, and positive collaboration mechanisms are either incomplete or inefficient. Strengthening these collaborations and ensuring that public resources are optimally distributed, that can be considered as essential steps in building urban resilience. Developing a coordinated emergency response framework that includes public resource allocation and integrated governance would help cities effectively prevent and manage complex, diverse risks (Feng et al. 2024; Liu et al. 2024; Xuesong et al. 2022).

In conclusion, enhancing urban resilience in developing cities requires addressing gaps in infrastructure, governance, data integration, and resource allocation. Prioritizing the construction of smart and resilient cities will not only improve a city's capacity to withstand disasters but also contribute to sustainable urban development, ensuring the safety and well-being of residents in times of crisis.

8.2.5 AI-Driven Scenario Planning for Developing Smart and Resilient Urban Environments

AI is increasingly seen as a promising solution to overcome the limitations of current scenario planning methods, playing a significant role in the development of smart and resilient cities (Hao et al. 2024). The integration of AI into urban planning has gained momentum due to its transformative potential (As et al. 2022; Andrews et al. 2022; Wasserman and Flaxman 2022; Sanchez et al. 2023; Hurtado et al. 2023). For example, planners anticipate that AI could take over repetitive and mundane tasks, such as traffic monitoring and sketch rendering, which are typically performed by entry-level planners (Hurtado et al. 2023; Ye et al. 2022; Zheng et al. 2023).

AI contributes to smart and resilient cities in at least two major ways when applied to plan generation. Firstly, it can produce comprehensive, data-driven, long-term plans that are not only methodologically robust but also extend beyond the conventional scope of human planners. This capability can lead to innovative approaches to urban challenges. Secondly, AI algorithms, such as reinforcement learning (RL), can be utilized for optimizing real-time decision-making regarding urban system operations in anticipation of events like tropical storms, thereby minimizing their adverse impacts (Hao et al. 2024).

While traditional scenario planning emphasizes long-term strategies to address future uncertainties, resilient urban planning also suggests flexible design and management practices to tackle foreseeable disruptions, such as an approaching storm (Wardekker et al. 2010). Various strategies have been proposed for achieving flexible planning, including the development of multifunctional spaces and the dynamic management of urban systems (Wardekker et al. 2010; Yildirim et al. 2021; Li and Wu 2022). However, the reconfiguration of space functions and the adaptive control of city facilities require complex optimization and rapid decision-making, areas where AI, particularly RL, can provide substantial support.

AI can enhance scenario construction and contribute to the growth of smart, resilient cities through several means. It can collaborate with planners to identify potential challenges that may be overlooked and generate complex scenarios to stress-test current urban planning and management practices. Furthermore, AI has the capability to create detailed, photorealistic depictions of adverse scenarios, facilitating better stakeholder engagement and evaluation (Hao et al. 2024).

Beyond spatial planning, generative deep learning models can simulate unexpected adverse scenarios to test proposed plans, offering proactive, resilient strategies for cities at risk. Researchers have employed such models to simulate spatiotemporal urban dynamics, including energy usage and traffic flow, providing valuable insights for refining existing urban plans and management systems (Dong et al. 2022; Wang et al. 2020).

8.2.5.1 Integrated Platform for Urban Planning and Scenario Analysis

While various AI techniques have been introduced for specific steps in scenario planning, a significant gap remains in the form of comprehensive, dedicated platforms that unify these methods for seamless deployment. Some leading technology companies exemplify progress in this field by offering a suite of AI-powered cloud-based products that facilitate urban and architectural design at multiple scales. Future enhancements could include integrating scenario analyses that consider uncertainties and adversities, as well as flexible connectivity with external AI plug-ins to support pre-investment decision-making for stakeholders (Xu et al. 2024; Hao et al. 2024).

8.2.5.2 Multimodal Generation and Metaverse Development for Planning Assessment

Integrating generative deep learning models with extended reality (XR) technologies offer a promising path for immersive, qualitative plan evaluation. While some advances have been made in this area, there is a need for comprehensive platforms that automate the full process, from AI-driven scenario generation to XR visualization and participatory plan evaluation. Additionally, AI can create multi-sensory experiences, such as synchronized visual and acoustic stimuli, for deeper participant engagement. This aspect remains underexplored but could be enhanced by combining metaverse technology with participatory planning. For example, a virtual version of a proposed community park could be constructed and experienced by residents who provide feedback, akin to beta testing in software development, before the real-life implementation (Ratican et al. 2023; Hao et al. 2024; Yang et al. 2024).

8.2.5.3 Urban Systems as Robotic Frameworks for Adaptive Decision-Making

Current Smart City initiatives often view cities as platforms for deploying various smart systems, like mobile apps. However, addressing future complexities, such as extreme events, necessitates an integrated approach where urban systems

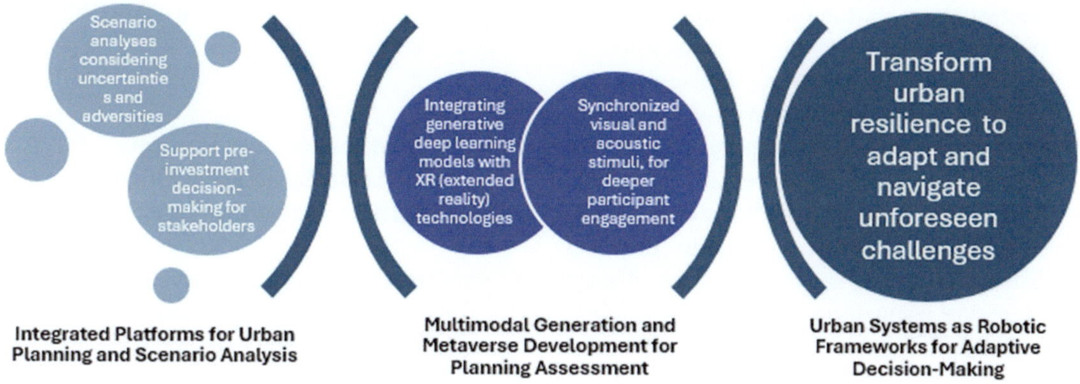

Fig. 8.3 AI-empowered scenario planning for building smart and resilient cities. (Source: Author's compilation)

collaborate dynamically. The concept of cities as large-scale robotic entities, where different systems interact and respond to both internal and external stimuli, could transform urban resilience. Incorporating deep reinforcement learning (RL)—a common approach in robotics—into urban systems would allow cities to adapt and navigate unforeseen challenges, much like an autonomous vehicle bypassing road obstacles, thereby enhancing resilience (Fan et al. 2023; Hao et al. 2024; Sun et al. 2024). AI-empowered scenario planning for building smart and resilient cities is illustrated in Fig. 8.3.

8.3 Framework of Land Spatial Resilience Planning and Construction

This section outlines the objectives and overarching principles of territorial spatial resilience planning and construction and develops a comprehensive framework encompassing the technical, spatial governance, and operational dimensions.

8.3.1 Objectives of Territorial Spatial Resilience Planning and Construction

The primary aim is to establish and enhance the national spatial resilience planning system, thereby boosting the adaptability, stability, and resilience of national territories while improving disaster prevention and mitigation capabilities. This goal supports economic growth, social stability, environmental protection, national security, and the well-being of citizens. The specific objectives include ensuring effective early warning systems before disasters, fostering calm and strategic responses during events, and promoting robust recovery and reconstruction efforts post-disaster. This involves developing goal-oriented spatial plans at various scales—micro, meso, and macro—anchored in "dual evaluation" and backed by emergency plans and supervisory structures (Assumma et al. 2024; Chen et al. 2023; Chioni et al. 2024; Li et al. 2024a; Rezvani et al. 2023; Shao et al. 2023; Wang et al. 2023a).

8.3.2 Core Concepts of Territorial Spatial Resilience Planning and Construction

Building resilient urban systems requires shifting from engineering-centric approaches to integrated governance, scenario-based planning, and multi-tiered spatial strategies that embed disaster resilience across all levels of territorial development.

1. Transition from Engineering to Governance for Enhanced Disaster Resilience: The safety framework for national and spatial disaster prevention must be reinforced with resilience to support post-disaster functionality. This

necessitates a shift in focus from purely engineering solutions to governance-based approaches, combining rigid planning with adaptive disaster prevention strategies to bolster overall resilience (Assumma et al. 2024; Chen et al. 2023; Chioni et al. 2024; Shao et al. 2023; Li et al. 2024a; Wang et al. 2023a).

2. Integration of Disaster Scenario Planning: Effective planning should incorporate disaster scenarios, using disaster risk assessments as a foundational guide for coordinating urban safety and resilience strategies. Theoretical models should leverage cross-disciplinary strengths, transitioning from urban to global, multi-scale approaches. These models should address the full timeline of disasters—from early warning to response and recovery—analyzing land space security mechanisms and functions under various stressors to provide a robust theoretical basis for emergency management and spatial protection (Mohammadi et al. 2023; Prashar et al. 2023; Han et al. 2024; Wang et al. 2023a)
3. Comprehensive Land and Space Disaster Prevention Planning: A multi-tiered disaster prevention planning system is essential. This system comprises (i) specialized plans for individual disaster types (e.g., flood control), (ii) targeted plans for disaster-related infrastructure (e.g., emergency evacuation routes), and (iii) integrated, multi-disaster prevention strategies that connect and support various plans, creating a holistic framework for territorial spatial disaster resilience (Hussain et al. 2023; Mao et al. 2024; Yang et al. 2023; Zhang et al. 2024a, b; Zhou et al. 2023; Wang et al. 2023a)
4. Embedding Resilience in Planning Frameworks: Establishing territorial spatial resilience requires a top-down design approach that clarifies the strategic framework, delineates tasks for each planning stage, and integrates the resilience concept throughout the entire process. This promotes synchronized development of resilient urban planning alongside broader territorial master plans, ensuring mutual reinforcement and closed-loop governance at the national level (Abujder Ochoa et al. 2024; Bühler et al. 2024; Williams 2023; Wang et al. 2023a).

8.3.3 Comprehensive Framework for Land Spatial Resilience Planning and Construction

To implement effective territorial spatial resilience planning, the concept must be embedded across all levels of spatial planning. This includes national-level regional plans, city or county-level plans integrating urban and rural areas, and township-level plans. These are aligned with comprehensive, detailed, and specialized plans. The proposed framework and technical route for resilience planning encompass:

1. Planning Structure and Coverage: Resilience planning must integrate special disaster prevention plans, foundational environmental plans, spatial structural plans, comprehensive development plans, and implementation mechanisms, forming a multidimensional and holistic resilience strategy (Banikoi et al. 2024; Vitale et al. 2023; Chen et al. 2023; Pascariu et al. 2023; Brand and Zonneveld 2024; Wang et al. 2023a).
2. "One Blueprint" Approach: The central focus is on the entire regional landscape and its elements, characterized by a hierarchy involving "five levels, four categories, and three systems," supporting various aspects of planning (Shao et al. 2023; Chen et al. 2023; Cui et al. 2024; Zhou et al. 2023; Fuggini et al. 2023; Wang et al. 2023a).
3. Planning Principles and Goals: The emphasis lies on pre-disaster early warning, in-disaster response, and post-disaster recovery, transforming the planning methodology, spatial focus, content, and strategy to establish a continuous, closed-loop governance model for spatial resilience planning (Azadi et al. 2023; Dong and Chen 2024; Ragazou et al. 2024; Williams 2023; Wang et al. 2023a).
4. Implementation and Technical Integration: Effective execution requires aligning technical methodologies, spatial governance frame-

works, and operational systems. The use of emerging technologies in combination with traditional disaster prevention strategies is vital. The planning process should begin with "dual evaluation," accounting for the entire disaster lifecycle—pre-event, event, and post-event—and quantifying relevant indicators to ensure an organized and systematic approach to disaster prevention and resilience (Ariyachandra and Wedawatta 2023; Assumma et al. 2024; Chen et al. 2023; Rezvani et al. 2023; Wang et al. 2023a).

8.4 Improvement Roadmap of Land Spatial Resilience Planning and Construction

This section summarizes a development pathway across these dimensions and proposes strategic measures in three key areas: enhancing the legal and regulatory framework, policy systems, and technical standards, to guide research and practical implementation in land space resilience planning.

8.4.1 Technical Method Level

8.4.1.1 Integrating Disaster Risk Assessment and "Dual Evaluation" for Disaster Planning

A crucial starting point for creating effective disaster prevention and mitigation plans is conducting a regional disaster risk assessment, which integrates with "dual evaluation" and "dual assessment" methods. "Dual evaluation" focuses on assessing the suitability of urban construction spaces and evaluating the carrying capacity of regional resources and the environment. However, while it covers the assessment of multiple overlapping risks, it often lacks an analysis of complex, cascading disaster scenarios triggered by major individual events. Additionally, challenges such as inconsistent data processing, varying indicators, and non-standardized evaluation systems make comprehensive assessments difficult. Therefore, multi-scale disaster risk assessment techniques are necessary for high, medium, and low-risk areas to identify high-risk zones effectively. These assessments should inform planning goals and provide a theoretical foundation for constructing resilient spatial frameworks, optimizing development patterns, and enhancing spatial governance (Wang et al. 2023a, b; Pratomo et al. 2024).

8.4.1.2 Leveraging Big Data for Enhanced Urban Governance

As urbanization accelerates and new technologies evolve, real-time big data analysis and visualization play an essential role in understanding urban dynamics. Big data allows for in-depth analysis and visualization of city activities and development, facilitating data-driven proposals for future planning. The use of big data also supports public participation, enabling residents to contribute insights and recommendations for urban development (Cao et al. 2019; Xu et al. 2023). Recent advancements in technology have revolutionized spatial planning, urban operations, and monitoring, promoting smart planning and informed decision-making processes. Digitalization has introduced an era of "smart planning" that is adaptive, perceptive, and conducive to green growth, aiding in building sustainable urban environments and fostering "smart land" initiatives globally (Wang et al. 2023a; Mehmood et al. 2024; Bibri et al. 2024a, b; Al-Mujahed 2024; Wang et al. 2023a).

8.4.1.3 Enhancing Disaster Management Technology with a Focus on Urban Areas

To promote efficient spatial planning, it is vital to align spatial planning efforts with comprehensive management standards across different levels. Urban centers, as key actors in disaster management, must prioritize technical approaches in urban planning. This begins with identifying and categorizing risk areas, strengthening disaster prevention and resilience in critical zones, and bolstering the management of urban infrastructure and public services. Emphasis should be placed on pre-disaster prevention, adaptability during disasters, and post-disaster recovery. Additionally, integrating disaster monitoring and assessment within urban evaluations, along with establishing a system of monitoring, response, analysis, and early warning, is essential. This integration ensures continuous improvement in spatial planning management and disaster response capabilities (Shan et al. 2022; Wang et al. 2023a; Watson et al. 2024; Xiong et al. 2022b).

8.4.2 Spatial Governance Level

8.4.2.1 Harmonizing Regulatory Constraints and Strategic Directives in Spatial Governance

Effective spatial governance involves balancing binding requirements, which safeguard public order and interests, and guiding requirements, which aim to optimize public welfare and foster high-quality development. When setting technical standards for planning, it is necessary to ensure adaptability to meet evolving spatial governance needs. Different control methods should be applied based on the area's characteristics: binding regulations for high-density development areas, stringent controls for ecological or cultural heritage sites, and flexible guiding requirements for economically beneficial zones. Establishing comprehensive technical standards for planning should respect market principles and the diverse needs of urban development while promoting systematic and sustainable land space growth (Grindsted et al. 2024; Koroso and Zevenbergen 2024; Nowak et al. 2023; Reinwald et al. 2024; Tan et al. 2024; Wang et al. 2023a).

8.4.2.2 Strengthening Governance Capacities in Critical Spatial Regions

Spatial governance capabilities include the effective coordination of land resources, optimization of spatial structures, and enhancement of disaster response to ensure public safety. To strengthen pre-disaster prevention, disaster adaptability, and post-disaster recovery, regional efforts should focus on disaster prevention infrastructure, engineering resilience, multi-hazard responses, and linked emergency mechanisms. Urban strategies should emphasize risk prevention for vital infrastructure and public facilities, with improved public service and living spaces to enhance flexibility (Amori 2024; Wang et al. 2023a).

8.4.3 Working System Level

8.4.3.1 Incorporating Resilience Goals into National Spatial Planning

National spatial planning occurs at multiple levels—national, provincial, municipal, county, and township—each with specific objectives and interlinked frameworks. Effective planning requires alignment and coordination across these levels, integrating project implementation with dynamic management mechanisms to form a closed-loop feedback system. National regional plans should emphasize risk zoning and strategic resilience goals, while provincial plans should focus on high-risk identification and cross-regional emergency strategies. Municipal and county-level plans should detail resilience measures, and township plans should strengthen the natural environment, infrastructure, and public services (Banikoi et al. 2024; Rezvani et al. 2023; Todella et al. 2024; Zhang et al. 2024a, b; Wang et al. 2023a).

8.4.3.2 Collaborative Multi-objective Spatial Planning

Territorial spatial planning should integrate special plans and coordinate elements to reflect multi-level, multi-objective needs. Collaborative planning connects departmental goals to ensure coherence and high-quality implementation. This approach involves synchronizing the resilience of urban living, production, and ecological spaces, avoiding fragmented improvements and maximizing resilience benefits. Moreover, aligning spatial resilience efforts with other strategic concepts—such as carbon neutrality, sponge cities, and green urban initiatives—ensures both immediate and long-term goals are met, preventing iso-

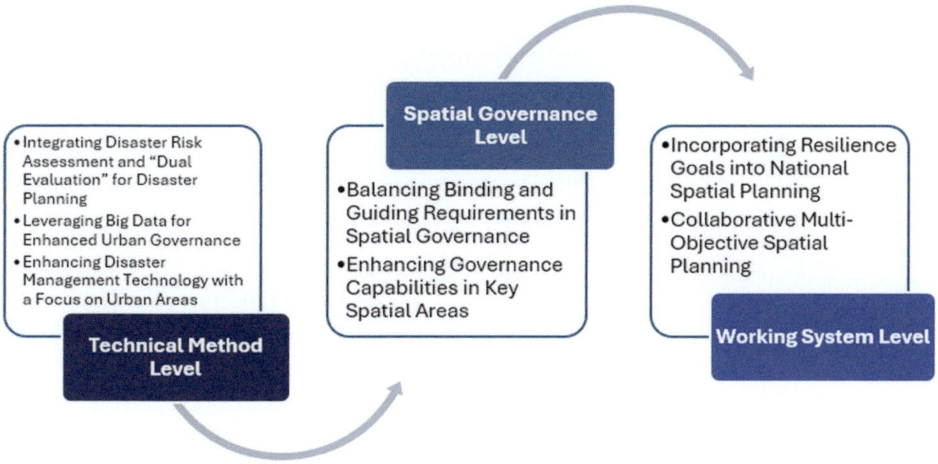

Fig. 8.4 Improvement roadmap of land spatial resilience planning and construction. (Source: Author's compilation)

lated resilience planning and fostering a unified approach (Mannucci 2024; Xue et al. 2024; Wang et al. 2023a, 2024). Figure 8.4 illustrates the improvement roadmap of land spatial resilience planning and construction.

8.5 Frameworks and Pathways for Developing Smart and Resilient Cities

8.5.1 Key Components of Smart and Resilient Urban Development

Building a smart resilient city involves a coordinated effort led by the government in collaboration with various social entities, all while integrating into the broader smart city development process. The foundation of a smart resilient city rests on several pillars: economic stability, social organization, well-maintained engineering facilities, advanced emergency management, information technology, and ecological and livable environments. Together, these elements enhance urban governance, improve risk prevention, and strengthen resilience from the bottom up (Houichi et al. 2024; Bibri et al. 2024a; Kaiser 2024; Khatibi et al. 2024; Xuesong et al. 2022). The construction framework for smart resilient cities is categorized into six levels:

- Smart Perception Layer: This layer integrates smart city technologies like the Internet of Things (IoT) and remote sensing to gather dynamic, real-time data. It creates a comprehensive understanding of urban conditions and provides the foundation for informed decision-making (Matei and Cocoșatu 2024; Xuesong et al. 2022).
- Resilience Perception Layer: Using data from the smart perception layer, the resilience perception layer models the effects of factors such as population growth, climate change, disaster outbreaks, and urban construction. These models assess the city's changing resilience capabilities (Mazzetto 2024; Xuesong et al. 2022).
- Resilience Modeling Layer: This layer involves algorithms such as big data, ML, and digital twins. It builds essential models like risk identification, monitoring, and loss assessment to create a resilient "urban brain." The models are updated regularly to ensure that the city's risk information is current and accurate (Alnaser et al. 2024; Bibri 2024; Fadhel et al. 2024; Xuesong et al. 2022).
- Resilience Decision-Making Layer: The decision-making framework covers the full cycle of urban risk management, from pre-disaster planning and assessment to emergency response and post-disaster recovery. This comprehensive system reduces potential

losses and improves the city's ability to respond to crises (Carramiñana et al. 2024; Xuesong et al. 2022).

- Resilience Synergy Layer: Divided into dimensions such as space, organization, institutions, technology, and culture, this layer promotes collaboration between various social entities. It enhances the city's internal structure and strengthens multi-dimensional coordination to handle emergencies (Bibri et al. 2024b; Xuesong et al. 2022).
- Resilience Management and Control Layer: This layer focuses on long-term stability by managing key urban sectors like transportation, healthcare, water supply, energy, and public services. It ensures that cities maintain normal operations even during disruptions (Li et al. 2024b; Razavi et al. 2024; Xuesong et al. 2022).

8.5.2 Ecological Framework for Smart and Resilient City Development

The smart and resilient city construction ecology emphasizes collaboration between the government, enterprises, and residents. This collaborative ecosystem is built around the concept of the resilient "urban brain," which combines smart technologies with resilience-building strategies. The system leverages the strengths of both government-led intelligence and enterprise innovation, fostering a multi-level network that integrates various fields and subjects (Shafik 2024; Wolniak and Stecuła 2024; Xuesong et al. 2022).

The "urban brain" functions as a central hub, comprising databases, functional assessment models, and decision-making algorithms that support urban emergency response. A key aspect of this system is the collaboration between industry, academia, and research institutions, which provides valuable feedback to governmental planning, policy development, and system innovation. Data sharing between the government and businesses is crucial for enhancing resilience, while residents are encouraged to participate actively in the emergency management process. This collective effort ensures that resources are mobilized efficiently during crises, and that grassroots-level emergency response is improved (Sivarethinamohan and Reddy 2024; Xuesong et al. 2022).

The smart and resilient city ecology operates on several guiding principles (Kangana et al. 2024; Oikonomaki et al. 2024):

- Collaboration Between Government and Enterprises: Enterprises bring technological expertise and operational experience to support the government's planning and standardization efforts.
- Crowdsourcing and Grassroots Innovation: Residents are empowered to contribute to emergency responses through crowdsourcing initiatives, allowing for real-time adaptation and local problem-solving.
- Long-term Cooperation: Establishing enduring partnerships between various stakeholders ensures continuous optimization of smart and resilient city systems, promoting sustainable urban development.

8.5.3 Pathways for Developing Smart and Resilient Cities

Constructing a smart and resilient city is a complex and iterative process. It requires prioritizing the development of basic, structural, and pre-set capabilities to avoid potential bottlenecks. By focusing on eight key processes, the path to building these cities involves enhancing six resilience capabilities, developing four urban governance functions, and creating a unified resilient "urban brain". The construction of a smart resilient city can be categorized into four primary areas (Firoozi and Firoozi 2024; Xuesong et al. 2022):

- Perception: This involves creating advanced systems to perceive and monitor changes in urban conditions, such as population shifts and infrastructure development.
- Cognition: This area focuses on understanding the implications of data collected through perception, utilizing algorithms and models to assess risks and inform decision-making.

- Thinking: The "urban brain" employs ML and other technologies to simulate potential outcomes of different scenarios, ensuring that the city can respond proactively to risks.
- Action: Smart cities must be able to act upon their insights in real time, mobilizing resources efficiently and implementing strategies to mitigate disaster impacts.

The resilient "urban brain" thus serves as the city's central decision-making hub. It integrates normal-time data collection with real-time crisis management, creating a flexible, adaptive system capable of handling both daily operations and emergency responses.

8.5.4 Evaluation Metrics for Smart and Resilient Cities

Assessing the effectiveness of a smart resilient city requires a comprehensive evaluation framework that combines both qualitative and quantitative measures. The goal of these assessments is to identify weaknesses in urban resilience, track improvements, and provide guidance for future development. The evaluation index system covers several core areas (Khatibi et al. 2024; Rezvani et al. 2024; Tsoutsa et al. 2024; Suárez et al. 2024; Xuesong et al. 2022):

- Organizational Management: This dimension focuses on the city's governance structures, leadership capabilities, and collaboration between different social entities. Indicators at this level assess how well a city can manage crises and implement modern management techniques.
- Economic and Social Factors: Economic resilience is critical for maintaining the city's long-term stability. Indicators in this category examine material security, flexible supply chains, and financial resilience. They also assess the city's ability to provide essential services, ensure social stability, and meet residents' basic survival needs during emergencies.
- Infrastructure: A resilient city's infrastructure must be able to withstand shocks and maintain operational functionality during crises. This includes transportation networks, energy supply, medical facilities, and environmental protection systems. Indicators focus on the quality and reliability of infrastructure services.
- Institutional Culture: Social and institutional resilience is shaped by public awareness, education, and collective identity. The construction of a resilient city depends not only on physical infrastructure but also on strong community ties, government support, and mutual cooperation among residents. This area measures how well residents and institutions are prepared to respond to emergencies and work together in times of crisis.

A well-rounded assessment system enables cities to understand their current resilience levels, identify gaps, and take strategic actions to improve overall disaster preparedness and urban governance. The continuous improvement of resilience indicators ensures that cities remain adaptive, secure, and capable of responding to both anticipated and unforeseen challenges.

In conclusion, smart resilient cities are the future of urban development, driven by the integration of technology, collaboration, and a focus on sustainability. From building a resilient "urban brain" to creating collaborative ecosystems and assessing resilience capabilities, smart cities are designed to be flexible, adaptive, and capable of withstanding a wide range of risks. By investing in infrastructure, strengthening governance, and fostering collaboration between the public and private sectors, cities can become more resilient and better equipped to protect their residents and ensure long-term prosperity (Khatibi et al. 2024; Al-Mujahed 2024; Rezvani et al. 2024; Das 2024; Abu-Rayash and Dincer 2023; Xuesong et al. 2022).

8.5.5 Strategic Recommendations for the Development of Smart and Resilient Cities

To effectively construct a smart and resilient city, it is essential to scientifically plan urban development. This involves transforming traditional urban concepts into a more distributed, gridded, and dynamic system that can adapt to sudden changes. The goal is to build a full-cycle, multi-level urban safety system that can resist impacts, adapt to disruptions, and recover quickly in the event of emergencies. The centralized urban construction model has proven vulnerable to large-scale disasters, so transitioning toward a more decentralized system is crucial (Hao et al. 2024; Mazzetto 2024; Al-Mujahed 2024; Zhou et al. 2023; Xuesong et al. 2022).

8.5.5.1 Enhancing Public Financial Investment and Infrastructure Development for Sustainable Growth

A key recommendation is to establish dedicated funds for the construction of smart and resilient cities. By coordinating the use of these funds, cities can ensure the effective allocation of resources for critical projects. This includes upgrading infrastructure and urban planning to support a more distributed and multi-center urban model, reducing the risks associated with overly concentrated city functions. Smart and resilient city construction also offers an opportunity to modernize existing infrastructure and implement a parallel urban transportation system with multiple options for mobility. Additionally, disaster prevention and emergency sites should be strategically located to ensure equal access to essential public services. This approach allows cities to maintain steady development, even in unpredictable environments (Attah et al. 2024; Alawi and Chu 2024; Ojo 2024; Meduri et al. 2024; Xuesong et al. 2022).

8.5.5.2 Strengthening Emergency Legislation and Social Governance Systems for Resilient Communities

The construction of smart and resilient cities should be elevated to a strategic priority, recognizing its role in ensuring the safety of national economic development. It is crucial to enhance policies and standards related to disaster reduction, emergency response, and post-disaster recovery. Local governments should regularly revise laws, regulations, and standards to reflect the evolving needs of disaster preparedness and recovery. Improving urban emergency management is vital. A responsive emergency network, combined with a flat management structure, ensures efficient action during crises. In addition, cities should implement digital security risk monitoring, early warning systems, and comprehensive disaster management information platforms. These technologies will improve risk prediction and enhance emergency response capabilities across various sectors (Lagap and Ghaffarian 2024; Kanbara et al. 2025; He and Faure 2024; Macatulad and Biljecki 2024; Xuesong et al. 2022).

8.5.5.3 Promote Technology-Empowered Development

The development of smart and resilient cities requires a gradual transition from informatization to intelligence and eventually to wisdom. Digital transformation is essential for urban governance, and the use of big data, cloud computing, and blockchain technology can fully empower cities. This transformation will improve digital government capabilities and enable the interconnection and sharing of data resources, supporting faster decision-making and crisis response. Building a resilient "urban brain" is a top priority. This system must integrate various data sources and eliminate "information islands" to create a cohesive and responsive smart city framework. Additionally, new infrastructure, such as big data centers, supercomputing centers, the Internet of Things, and industrial and satellite internet, must be developed to support intelligent city construction. Establishing an open platform for technology exchange can promote innovation in smart city development. This will create a scalable and replicable model for building smart and resilient cities, moving from pilot programs to broader regional implementations (Houichi et al. 2024; Attah et al. 2024; Zhuang et al. 2024; Mazzetto 2024; Matei and Cocoşatu 2024; Xuesong et al. 2022).

8.5.5.4 Encourage Government Coordination and Diversified Collaboration

Effective urban governance requires a collaborative approach that brings together government, enterprises, community residents, and social organizations. The principle of co-construction, co-governance, and sharing should guide the development of a resilient urban governance model for the new era. The government should play a central role in coordinating efforts across sectors. This includes integrating national and local governance, aligning peacetime and wartime responses, and fostering collaboration between public and private sectors. Grassroots organizations can help engage urban residents in the governance process, allowing them to contribute ideas and actively participate in building smart and resilient cities. Social forces, such as high-tech enterprises and professional institutions, should also be encouraged to participate in smart city construction. By fostering government-enterprise cooperation and creating a multi-win environment, cities can benefit from shared progress, openness, and collaboration, which is essential for sustainable urban development. The successful development of smart and resilient cities hinges on a combination of financial investment, legal frameworks, technological innovation, and collaborative governance. By focusing on infrastructure improvements, emergency preparedness, and technology integration, cities can build a resilient foundation capable of withstanding the challenges of the future. Moreover, fostering partnerships between the government, private enterprises, and residents will create a new urban development ecosystem that promotes long-term stability, security, and progress (Bradley and Mahmoud 2024; Martín-Rojo 2024; Mulyawati et al. 2024; Nichi and Cortese 2024; Xuesong et al. 2022). Figure 8.5 illustrates the framework and pathways for building smart and resilient cities.

8.6 Next Level of Smart and Resilient City: Concept of Smart-Resilient 5.0

Smart-Resilient 5.0 leverages technological innovations and digitalization to redefine urban governance and economic activities, aiming to

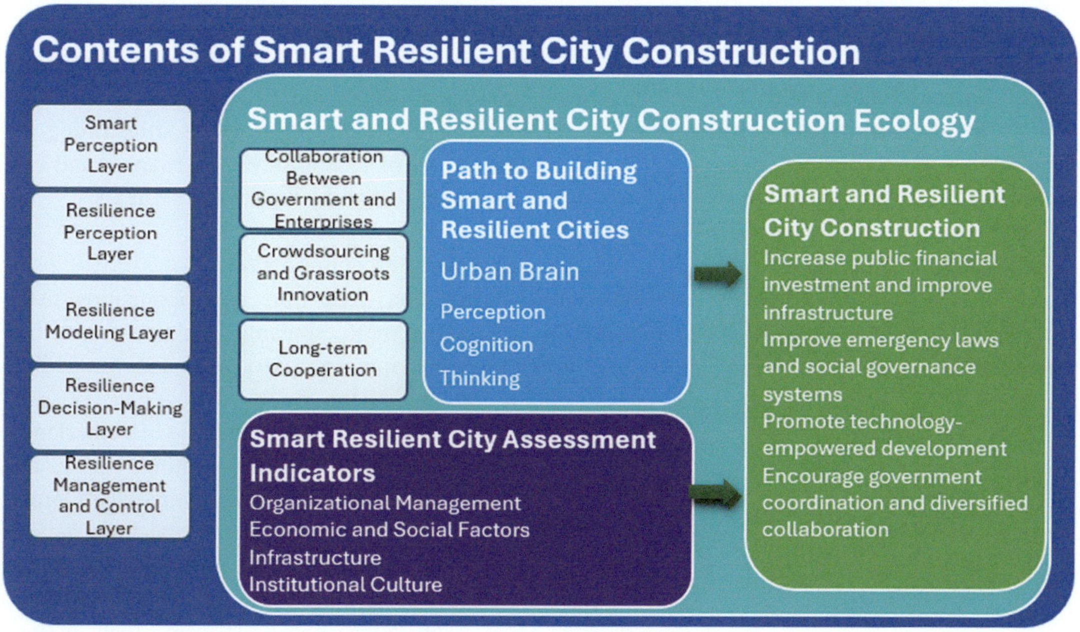

Fig. 8.5 Framework and pathways for building smart and resilient cities. (Source: Author's compilation)

deliver sustainable, inclusive, and well-managed societal outcomes in smart and resilient cities (Hunter et al. 2022). This concept extends the smart city framework by integrating sustainability, inclusivity, and resilience into urban development (Rosemann et al. 2021; Becker et al. 2023). In the face of rapid urbanization, Smart-Resilient 5.0 emerges as a transformative vision that aligns technology, innovation, and sustainability to address modern urban challenges (Scribano 2021; Yun and Lee 2019; Boumale and Tamine 2022). It focuses on improving resource efficiency, enhancing connectivity, and embedding data-driven governance as a foundation for addressing urban problems (Fernandez-Anez et al. 2018; Lin et al. 2023; Wolniak et al. 2023). Incorporating technologies such as IoT, AI, and big data, Smart-Resilient 5.0 transforms urban environments into smarter, more efficient, and sustainable spaces (Yigitcanlar et al. 2018). While the smart city emphasizes improving individual aspects of urban life, Smart-Resilient 5.0 broadens this focus by fostering an integrated technological ecosystem that ensures intelligent connectivity and advanced governance, representing the evolution of smart city frameworks (Fernandez-Anez et al. 2018; Zubizarreta et al. 2016).

8.6.1 Smart-Resilient 5.0: Environmental Perspective

Smart-Resilient 5.0 emphasizes sustainability, aligning with the UN's Sustainable Development Goals (SDGs), which focus on environmental, social, and economic well-being (Kamruzzaman et al. 2015; Grossi and Trunova 2021). Key components include:

Technological Integration: AI and big data optimize urban functions like transportation, energy consumption, and waste management to reduce environmental impacts (Singh et al. 2020).

Smart Infrastructure: Systems for renewable energy, waste reduction, and intelligent transportation promote sustainable resource management while enhancing urban livability (Svítek et al. 2019; Selim et al. 2018).

Data-Driven Governance: Collecting and analyzing data facilitates efficient urban planning, ensuring eco-conscious decision-making (Shahat-Osman and Elragal 2021). Engaging communities and fostering transparency further enhance governance (Almalki et al. 2023).

Environmental Benefits: Smart-Resilient 5.0 mitigates environmental degradation through green spaces, improved air quality, biodiversity conservation, and climate resilience (Angelidou et al. 2018; Makieła et al. 2022).

Social Equity: Balancing technological progress with social inclusion, Smart-Resilient 5.0 addresses housing affordability, digital access, and community engagement (Esmaeilpoorarabi et al. 2018; Chen et al. 2022).

Environmental perspective of Smart-Resilient 5.0 Concept is illustrated in Fig. 8.6.

8.6.2 Smart-Resilient 5.0 City Concept: Use Cases by Region and Foundational Layers

This section explores how leading cities across Europe, the UK, the US, China, and North Cyprus implement smart and resilient urban strategies through six foundational layers, illustrating the integration of technology, governance, and collaboration in diverse real-world contexts.

(a) Smart Perception Layer
Function: Real-time data acquisition using IoT, remote sensing, and environmental monitoring for informed and immediate urban decision-making.
- Europe—Amsterdam, Netherlands: Amsterdam's smart lighting, traffic, and waste sensors (under Amsterdam Smart City initiative) collect real-time data to optimize energy use and traffic flows.
- UK—London: The London Datastore is an open data platform that gathers IoT-sourced data on mobility, pollution,

Fig. 8.6 Environmental perspective of Smart-Resilient 5.0 concept. (Source: Author's compilation)

and service use for public and institutional access.

- US—New York City: NYC's LinkNYC kiosks and connected infrastructure continuously feed data into urban dashboards to monitor pedestrian traffic and environmental conditions.
- China—Hangzhou: Alibaba's City Brain project collects data from cameras, sensors, and signals to optimize traffic light control and emergency vehicle prioritization.
- North Cyprus—Girne/Catalköy-Esentepe (pilot areas): In emerging projects, IoT-enabled smart meters and water flow sensors are being proposed for energy and utility monitoring in tourist-heavy municipalities.

(b) Resilience Perception Layer
Function: Identification and modeling of systemic urban vulnerabilities (e.g., climate risks, demographic changes) to understand stress points.

- Europe—Copenhagen, Denmark: Uses environmental sensors and climate models to assess flood risks, integrating green infrastructure into vulnerable zones (e.g., cloudburst plans).
- UK—Glasgow: Under its Future City Programme, the city developed models of vulnerable populations and infrastructure stressors tied to energy poverty and weather extremes.
- US—San Francisco: The city maps earthquake-prone zones and uses digital infrastructure to assess building vulnerabilities and heat island effects.
- China—Shenzhen: With AI-driven demographic analysis, Shenzhen models population surges, density hotspots, and healthcare access gaps for resilience planning.
- North Cyprus—Catalköy–Esentepe: Rapid tourism-driven population growth and unplanned coastal development are key vulnerabilities under initial spatial stress assessments.

(c) Resilience Modeling Layer
Function: Applying AI, machine learning, and digital twin technologies to simulate disasters, test interventions, and forecast systemic disruptions.

- Europe—Helsinki, Finland: Uses digital twins for energy efficiency modeling, traffic simulations, and building retrofitting scenarios to support long-term sustainability planning.
- UK—Manchester: The city's digital twin initiative simulates the effects of extreme weather events on infrastructure and helps prioritize investment in social housing resilience.
- US—Boston: Digital twin modeling supports sea-level rise preparedness and simulates urban heat effects for targeted greening.
- China—Shanghai: Operates a fully integrated city digital twin for simulating fire outbreaks, industrial accidents, and crowd management in megastructures.
- North Cyprus—Catalköy–Esentepe (proposed): Planned deployment of digital twin infrastructure in collaboration with planning agencies to simulate tourist-season pressure, coastal erosion, and water network strain.

(d) Resilience Decision-Making Layer
Function: Development of multi-hazard risk governance systems from preparedness to recovery, guided by real-time modeling outputs.

- Europe—Milan, Italy: Operates a city-wide decision-support platform integrating climate models, risk indicators, and real-time data to inform zoning and public alerts.
- UK—Bristol: Combines data from health, mobility, and environment into dashboards to inform decision-makers during emergencies.
- US—Los Angeles: Utilizes predictive policing and real-time risk analytics to anticipate urban disturbances and deploy coordinated responses.
- China—Beijing: Advanced smart grids and emergency response platforms use predictive analytics for energy and air quality crises.
- North Cyprus—Civil Defense & Municipal Coordination: Currently transitioning from reactive protocols to integrated resilience planning, using GIS-linked early warning systems.

(e) Resilience Synergy Layer
Function: Cross-sectoral and intergovernmental collaboration in resilience building—uniting public services, private innovation, and civil society action.

- Europe—Vienna, Austria: A model of co-creation and collaborative governance, Vienna's Smart City Framework involves housing cooperatives, NGOs, and universities in climate resilience planning.
- UK—Birmingham: Engages community groups and local businesses in the Birmingham Resilience Strategy, particularly in food security and flood protection.
- US—Portland, Oregon: Leverages public-private partnerships (PPPs) and grassroots networks to co-develop green infrastructure for resilience.
- China—Chengdu: Integrates resilience into spatial planning by involving regional agencies and industry consortia.
- North Cyprus—Catalköy–Esentepe: Emerging efforts to engage local civil society (e.g., CADER) and SMEs in co-creating water efficiency and tourism sustainability plans.

(f) Resilience Management and Control Layer
Function: Ensures continuity and adaptive stability of core urban systems (energy, transport, healthcare, utilities) during shocks or prolonged stress.

- Europe—Stockholm, Sweden: Operates citywide microgrids and transport systems designed to remain functional during extreme weather or cyberattacks.
- UK—Leeds: Implements smart grid technologies and localized emergency power backups for hospitals and schools.

- US—Houston: Post-Hurricane Harvey, the city has deployed resilient stormwater infrastructure and hospital system redundancies.
- China—Wuhan: Post-COVID-19, Wuhan invested in real-time hospital inventory systems, autonomous health supply chains, and integrated mobility planning for crisis situations.
- North Cyprus—Girne/Catalköy: Currently addressing weaknesses in power grid stability, emergency medical coordination, and unconnected transport systems through proposed infrastructure resilience upgrades.

The construction of smart and resilient cities is not merely a technological endeavor—it is a layered, adaptive process requiring aligned actions across governance, engineering, and community systems. From IoT-enabled sensing to digital twin-powered decision-making, each layer builds toward a robust "urban brain" capable of thinking, sensing, and acting under uncertainty. The future of urban sustainability lies in the synthesis of perception, cognition, thinking, and action—as cities move toward inclusive, data-informed, and risk-resilient systems.

8.6.3 Prominent Research Areas in Smart-Resilient 5.0

Smart-Resilient 5.0 highlights the importance of global collaboration in addressing urbanization and climate challenges. By sharing best practices and fostering innovation, cities can jointly navigate the complexities of sustainable urban growth (Wilson and Daugherty 2018).

Key research areas include:

Collaborative Intelligence: Integrating human expertise with AI to solve urban challenges and promote responsible innovation through initiatives like hackathons (Batty 2013).

Education and Digital Inclusion: Expanding digital networks and ensuring cybersecurity while addressing societal and governance issues (Barns et al. 2017; Ahmad et al. 2021).

Smart-Resilient 5.0 represents a global strategy for creating smarter, more sustainable cities, balancing technological advancements with societal and environmental priorities.

8.7 Conclusion

8.7.1 Advancing Smart and Resilient Cities

The development of smart and resilient cities represents a transformative approach to urban planning and governance, addressing the growing challenges of modern urbanization. These cities integrate cutting-edge technologies, such as AI, IoT, and big data, with strategies for disaster resilience to create urban environments that are not only efficient and adaptive under normal circumstances but also robust in the face of crises. The concept of a smart resilient city bridges the gap between traditional smart cities and disaster-resilient urban frameworks, fostering seamless operations and rapid recovery.

At the core of this approach is the combination of technological innovation and resilience-building measures. Smart cities contribute data-driven tools, real-time analytics, and connected infrastructures, while resilient cities enhance governance structures, emergency preparedness, and recovery capabilities. Together, they create a dynamic urban model capable of proactive risk assessment, effective disaster response, and long-term adaptability. This integrated model goes beyond post-disaster recovery by emphasizing early warning systems, predictive decision-making, and continuous learning from past events.

The significance of smart resilient cities lies in their ability to address critical needs in disaster prevention, urban governance, and resource allocation. They not only safeguard public safety during emergencies but also improve the quality of life for citizens by ensuring efficient public services, sustainable resource management, and equitable access to infrastructure. Furthermore, the integration of AI into scenario planning enhances the adaptability of urban systems, enabling cities to predict and mitigate the impacts of natural and human-made disasters more effectively.

In conclusion, smart and resilient cities offer a comprehensive, technology-driven framework for addressing the complexities of urbanization. By combining intelligence and resilience, they pave the way for a sustainable, inclusive, and adaptive future, transforming urban areas into hubs of innovation and security for generations to come.

8.7.2 Framework and Roadmap for Territorial Spatial Resilience Planning

The framework for territorial spatial resilience planning and construction provides a comprehensive approach to strengthening the adaptability and robustness of urban and regional landscapes. It emphasizes three key dimensions: technical methodologies, spatial governance, and operational systems, ensuring that resilience becomes an integral part of land and urban planning processes.

The primary objectives of resilience planning include the development of national and regional systems that can effectively prevent, respond to, and recover from disasters. By leveraging goal-oriented plans at micro, meso, and macro levels, this approach aims to improve disaster preparedness, foster stability during crises, and facilitate post-disaster recovery. These plans integrate advanced risk assessment methodologies and "dual evaluation" techniques to identify high-risk areas, enabling targeted disaster prevention and mitigation strategies.

A significant transition is proposed, shifting from traditional engineering-based disaster prevention to governance-based resilience frameworks. This change combines rigid infrastructural planning with flexible and adaptive strategies, addressing complex and cascading disaster scenarios. The inclusion of disaster scenario planning in spatial resilience strategies provides a multi-scale perspective, offering robust theoretical and practical foundations for urban and territorial emergency management.

The "one blueprint" approach ensures holistic planning by harmonizing multiple dimensions of spatial planning—disaster prevention, environmental management, and development strategies—into a unified system. At the technical level, emerging technologies such as big data analytics and real-time monitoring are integrated to improve decision-making, public participation, and disaster management. Spatial governance is enhanced through a balance of binding regulations and adaptive guiding requirements, optimizing resource allocation and ensuring sustainable development.

This roadmap also emphasizes collaborative multi-objective planning, ensuring coherence across resilience, ecological, and economic goals. Aligning resilience strategies with broader initiatives like carbon neutrality and green urban development fosters long-term sustainability. Ultimately, this framework provides a structured and dynamic approach to territorial resilience, safeguarding public safety, enhancing urban functionality, and supporting sustainable development in the face of increasing risks and uncertainties.

8.7.3 Framework and Pathways for Smart and Resilient Cities

The development of smart and resilient cities hinges on a robust framework that integrates technological innovation, governance structures, and social collaboration to address the challenges of urbanization and climate change. This framework encompasses multiple layers, from real-time data collection to long-term resilience strategies, creating a cohesive and adaptive urban ecosystem.

The construction of a smart resilient city involves six foundational layers:

(a) Smart Perception Layer: Leveraging technologies like IoT and remote sensing to gather real-time urban data, enabling informed decision-making.
(b) Resilience Perception Layer: Modeling urban vulnerabilities, such as population

growth and climate change, to assess resilience capabilities.
(c) Resilience Modeling Layer: Employing algorithms, big data, and digital twins to build models for risk assessment and disaster monitoring.
(d) Resilience Decision-Making Layer: Providing a comprehensive risk management system, spanning pre-disaster planning to post-disaster recovery.
(e) Resilience Synergy Layer: Promoting collaboration across spatial, organizational, and cultural dimensions to enhance emergency response.
(f) Resilience Management and Control Layer: Ensuring the stability of critical urban sectors like transportation, energy, and public services during crises.

The ecology of smart and resilient cities thrives on partnerships between government, enterprises, and residents. This collaboration integrates technological advancements with resilience strategies to create a unified "urban brain." Government-led initiatives align with enterprise innovation and community participation, fostering a multi-level network that supports effective resource allocation and emergency response.

The path to building these cities emphasizes four key areas: perception, cognition, thinking, and action. These processes ensure cities can monitor, analyze, and act upon insights in real-time, leveraging advanced technologies like AI and big data. Comprehensive assessment indicators further guide cities in tracking progress, identifying weaknesses, and improving urban governance.

In conclusion, smart resilient cities represent the future of urban development, blending technological advancements with sustainability and inclusivity to create adaptive and sustainable environments. Through collaboration, innovation, and governance, these cities ensure resilience against future challenges while enhancing quality of life.

References

Abujder Ochoa WA, Iarozinski Neto A, Vitorio Junior PC, Calabokis OP, Ballesteros-Ballesteros V (2024) The theory of complexity and sustainable urban development: a systematic literature review. Sustainability 17(1):3

Abu-Rayash A, Dincer I (2023) Development of an integrated sustainability model for resilient cities featuring energy, environmental, social, governance and pandemic domains. Sustainable Cities and Society 92:104439

Ahmad N, Laplante P, DeFranco J, Kassab M (2021) A cybersecurity educated community. IEEE Trans Emerg Top Comput 10:1456–1463

Alawi M, Chu D (2024) New approach to integrating disaster resilience into public open space planning and design. J Infrastruct Policy Dev 8(9):5222

Ali M, Naeem F, Adam N, Kaddoum G, Adnan M, Tariq M (2023) Integration of data driven technologies in smart grids for resilient and sustainable smart cities: a comprehensive review. arXiv preprint arXiv:2301.08814

Almalki F, Alsamhi S, Sahal R, Hassan J, Hawbani A, Rajput N, Saif A, Morgan J, Breslin J (2023) Green IoT for eco-friendly and sustainable smart cities: future directions and opportunities. Mob Netw Appl 28:178–202

Al-Mujahed LKA (2024) Smart cities and sustainable development goals: a systematic review and future research directions. J Sustain Econ 1(1)

Almulhim AI (2025) Building urban resilience through smart city planning: a systematic literature review. Smart Cities 8(1):22

Alnaser AA, Maxi M, Elmousalami H (2024) AI-powered digital twins and internet of things for smart cities and sustainable building environment. Appl Sci 14(24):12056

Amori SLE (2024) 11 creating adaptive smart cities. Sustainable Urbanization in the Caribbean.

Andrews C, Cooke K, Gomez A, Hurtado P, Sanchez T, Shah S, Wright N (2022) AI in planning: opportunities and challenges and how to prepare. American Planning Association, Chicago

Angelidou M, Psaltoglou A, Komninos N, Kakderi C, Tsarchopoulos P, Panori A (2018) Enhancing sustainable urban development through smart city applications. J Sci Technol Policy Manag 9:146–169

Ariyachandra MMF, Wedawatta G (2023) Digital twin smart cities for disaster risk management: a review of evolving concepts. Sustainability 15(15):11910

As I, Basu P, Talwar P (2022) Artificial intelligence in urban planning and design: technologies, implementation, and impacts. Elsevier, Amsterdam

Assumma V, Bottero M, Ishizaka A (2024) Enhancing territorial resilience assessment with a decision-aiding model in regional planning of Socio-Ecological Systems. Environmental Science & Policy 154:103691

Attah RU, Gil-Ozoudeh I, Garba BMP, Iwuanyanwu O (2024) Leveraging geographic information systems and data analytics for enhanced public sector decision-making and urban planning. Magna Sci Adv Res Rev 12(2):152–163

Azadi H, Robinson G, Barati AA, Goli I, Moghaddam SM, Siamian N et al (2023) Smart land governance: towards a conceptual framework. Land 12(3):600

Banikoi H, Dakyaga F, Cobbinah PB (2024) Embedding spatial planning in contemporary multi-level governance: the sustainability entanglement. Land Use Policy 146:107324

Barns S, Cosgrave E, Acuto M, Mcneill D (2017) Digital infrastructures and urban governance. Urban Policy Res 35:20–31

Batty M (2013) Big data, smart cities and city planning. Dialogues Hum Geogr 3:274–279

Becker J, Chasin F, Rosemann M, Beverungen D, Priefer J, Brocke J, Matzner M, del Rio Ortega A, Resinas M, Santoro F et al (2023) City 5.0: citizen involvement in the design of future cities. Electron Mark 33:10

Bibri SE (2024) Artificial intelligence of things for smarter eco-cities: pioneering the environmental synergies of urban brain, digital twin, metabolic circularity, and platform. CRC Press

Bibri SE, Huang J, Jagatheesaperumal SK, Krogstie J (2024a) The synergistic interplay of artificial intelligence and digital twin in environmentally planning sustainable smart cities: a comprehensive systematic review. Environ Sci Ecotechnol 100433

Bibri SE, Huang J, Krogstie J (2024b) Artificial intelligence of things for synergizing smarter eco-city brain, metabolism, and platform: pioneering data-driven environmental governance. Sustain Cities Soc 108:105516

Boumale B, Tamine R (2022) Synergy between smart cities and industry 4.0 in public spaces: bibliometric analysis. Procedia Comput Sci 204:775–783

Bradley S, Mahmoud IH (2024) Strategies for co-creation and co-governance in urban contexts: building trust in local communities with limited social structures. Urban Sci 8(1):9

Brand N, Zonneveld W (2024) Water and spatial planning in the Netherlands: the latent potential of spatial planning for flood resilience. In: Oxford research encyclopedia of environmental science

Brandt S, Siebert J (2022) Challenges related to system-of-systems for greening and climate adaptation in smart cities. arXiv preprint arXiv:2211.08890

Bühler M, Hollenbach P, Köhler L, Armstrong R (2024) Unlocking resilience and sustainability with earth-based materials: a principled framework for urban transformation. Front Built Environ 10:1385116

Cao Y, Zhen F, Xi GL (2019) The intelligent urban governance based on big data: international experience and China strategy. Urban Plann Int 34(3):71–77

Carramiñana D, Bernardos AM, Besada JA, Casar JR (2024) Towards resilient cities: a hybrid simulation framework for risk mitigation through data-driven decision making. Simul Model Pract Theory 133:102924

Chakraborty A, McMillan A (2015) Scenario planning for urban planners: toward a practitioner's guide. J Am Plan Assoc 81(1):18–29

Chen T, Shi CM, Wang GY (2021) Research on urban water environment resilience planning under the background of climate change: a case study of Singapore. Urban Plann Int 36(5):52–60

Chen T, Ramon Gil-Garcia J, Gasco-Hernandez M (2022) Understanding social sustainability for smart cities: the importance of inclusion, equity, and citizen participation as both inputs and long-term outcomes. J Smart Cities Soc 1:135–148

Chen S, Zhu C, Ouyang X, Han Y (2023) Research on the supervision and implementation system of territorial space planning from the perspective of resilience. Sustainability 15(20):14682

Chioni C, Pezzica C, Favargiotti S (2024) Territorial digital twins: a key for increasing the community resilience of fragile mountain inner territories? Sustain Dev 32(2):1548–1563

Cui J, Jin H, Kong X, Sun J, Peng Y, Zhu Y (2024) Territorial spatial resilience assessment and its optimization path: a case study of the Yangtze River Economic Belt, China. Land 13(9):1395

Das DK (2024) Exploring the symbiotic relationship between digital transformation, infrastructure, service delivery, and governance for smart sustainable cities. Smart Cities 7(2):806–835

Dong W, Chen X (2024) Evaluation of community regeneration planning and resilience based on fuzzy group decision-making method. J Urban Plann Dev 150(3):04024024

Dong W, Chen X, Yang Q (2022) Data-driven scenario generation of renewable energy production based on controllable generative adversarial networks with interpretability. Appl Energy 308:118387

Esmaeilpoorarabi N, Yigitcanlar T, Guaralda M, Kamruzzaman M (2018) Evaluating place quality in innovation districts: a delphic hierarchy process approach. Land Use Policy 76:471–486

Fadhel MA, Duhaim AM, Saihood A, Sewify A, Al-Hamadani MN, Albahri AS, Gu Y (2024) Comprehensive systematic review of information fusion methods in smart cities and urban environments. Inf Fusion 102317

Fan X, Zhang X, Wang X, Yu X (2023) A deep reinforcement learning model for resilient road network recovery under earthquake or flooding hazards. J Infrastruct Preserv Resil 4(1):8

Feng Y, Wang J, Zhang T (2024) The impact of smart city policies on city resilience: an evaluation of 282 Chinese cities. Sustainability 16(19):8669

Fernandez-Anez V, Fernández-Güell J, Giffinger R (2018) Smart city implementation and discourses: an integrated conceptual model: the case of Vienna. Cities 78:4–16

Firoozi AA, Firoozi AA (2024) Neuromorphic computing: transforming disaster management and resilience in civil engineering. Springer Nature

Fuggini C, Solari C, De Stefano R, Bolletta F, De Maio FV (2023) Assessing resilience at different scales: from single assets to complex systems. Environ Syst Decis 43(4):693–707

Grindsted TS, Larsen MT, Jensen HT (2024) Strategic planning toward sustainable urban planning models? The case of municipal directors in Denmark. In: Geoethics for the Future. Elsevier, pp 285–298

Grossi G, Trunova O (2021) Are UN SDGs useful for capturing multiple values of smart city? Cities 114:103193

Han F, Yu J, Zhou G, Li S, Sun T (2024) Projected urban flood risk assessment under climate change and urbanization based on an optimized multi-scale geographically weighted regression. Sustain Cities Soc 112:105642

Hao H, Wang Y, Chen J (2024) Empowering scenario planning with artificial intelligence: a perspective on building smart and resilient cities. Engineering. https://doi.org/10.1016/j.eng.2024.06.012

He Q, Faure M (2024) Strengthening resilience and sustainability for post-disaster recovery: a comparative law and economics analysis on smart mixes between mechanisms. Sustainability 16(21):9534

Houichi M, Jaidi F, Bouhoula A (2024) A comprehensive and in-depth study of the threats faced by smart cities and the countermeasures implemented in their key areas. J Infrastruct Policy Dev 8(10):8629

Hunter M, Soro A, Brown R, Harman J, Yigitcanlar T (2022) Augmenting community engagement in city 4.0: considerations for digital agency in urban public space. Sustainability 14:9803

Hurtado P, Shah S, DeAngelis J, Gomez A (2023) 2023 trend report for planners. American Planning Association, Chicago

Hussain MA, Shuai Z, Moawwez MA, Umar T, Iqbal MR, Kamran M, Muneer M (2023) A review of spatial variations of multiple natural hazards and risk management strategies in Pakistan. Water 15(3):407

Kaiser ZA (2024) Smart governance for smart cities and nations. J Econ Technol 2:216–234

Kamruzzaman M, Hine J, Yigitcanlar T (2015) Investigating the link between carbon dioxide emissions and transport-related social exclusion in rural Northern Ireland. Int J Environ Sci Technol 12:3463–3478

Kanbara S, Shaw R, Eguchi K, Das S (2025) Lessons from the 2024 Noto Peninsula Earthquake: need for digital transformation in disaster response. Progress Disaster Sci 25:100400

Kangana N, Kankanamge N, De Silva C, Goonetilleke A, Mahamood R, Ranasinghe D (2024) Bridging community engagement and technological innovation for creating smart and resilient cities: a systematic literature review. Smart Cities 7(6):3823–3852

Khatibi H, Wilkinson S, Sweya LN, Baghersad M, Dianat H (2024) Navigating climate change challenges through smart resilient cities: a comprehensive assessment framework. Land 13(3):266

Koroso NH, Zevenbergen JA (2024) Urban land management under rapid urbanization: exploring the link between urban land policies and urban land use efficiency in Ethiopia. Cities 153:105269

Lagap U, Ghaffarian S (2024) Digital post-disaster risk management twinning: a review and improved conceptual framework. Int J Disaster Risk Reduct 110:104629

Leitner S (2023) Building resilient organizations: the roles of top-down vs. bottom-up organizing. arXiv preprint arXiv:2305.07352

Li S, Wu T (2022) Deep reinforcement learning-based decision support system for transportation infrastructure management under hurricane events. Struct Saf 99:102254

Li R, Di Y, Tian H (2024a) Computer-aided resilience: advanced techniques for disaster management in smart urban environments. Sustain Cities Soc 108:105437

Li J, Zou Y, Li J, Si S (2024b) Urban spatial, engineering and management resilience capacity enhancement from a flood safety perspective. Water-Energy Nexus

Lin Y, Hu W, Chen X, Li S, Wang F (2023) City 5.0: towards spatial symbiotic intelligence via DAOs and parallel systems. IEEE Trans Intell Veh 8:3767–3770

Liu S, Wu Y, Jiang G, Dong J (2024) System dynamics modeling of collaborative governance in smart cities: a case study of Dongguan, China. Sci Rep 14(1):31758

Long Ł, Junjia Y, Jiao W, Alias AH, Haron NA, Abu Bakar N (2025) Urban flood resilience evaluation in China: a systematic review of frameworks, methods, and limitations. Geomat Nat Haz Risk 16(1):2445631

Macatulad E, Biljecki F (2024) Continuing from the Sendai Framework midterm: opportunities for urban digital twins in disaster risk management. Int J Disaster Risk Reduct 102:104310

Makieła Z, Stuss M, Mucha-Kuś K, Kinelski G, Budziński M, Michałek J (2022) Smart city 4.0: sustainable urban development in the Metropolis GZM. Sustainability 14:3516

Mannucci S (2024) Climate adaptation in urban planning: toward sustainable and resilient urban environments. Springer Nature

Mao Y, Zhuo J, Zhai D, Zhang J (2024) Reflection and reconstruction of logical framework of comprehensive transportation planning in the territorial spatial planning system from the perspective of resilience. In: CICTP 2024, pp 2726–2734)

Marin V, Molinero C, Arcaute E (2024) The scalar mismatch of regional governance: a comparative analysis of hierarchical structures. Environ Plann B Urban Analyt City Sci 51:23998083241234405

Martín-Rojo I (2024) Strategic planning for a smart sustainable city model: the importance of public administration and enterprises cooperation. In: The Strategic paradigm of csr and sustainability: a public-private approach. Cham, Springer Nature Switzerland, pp 183–203

Matei A, Cocoşatu M (2024) Artificial Internet of Things, sensor-based digital twin urban computing vision algorithms, and blockchain cloud networks in sustainable smart city administration. Sustainability 16(16):6749

Mazzetto S (2024) A review of urban digital twins integration, challenges, and future directions in smart city development. Sustainability 16(19):8337

Meduri Y, Singh R, Manoharan G (2024) Key networks to create disaster resilient Smart Cities Mission: a case for remodeling India's Smart Cities Mission to include disaster resilience. Resil Cities Struct 3(4):52–62

Mehmood R, Yigitcanlar T, Corchado JM (2024) Smart technologies for sustainable urban and regional development. Sustainability 16(3):1171

Meng FK, Wu XL (2022) Revisiting "smart city": three basic research questions—based on a systematic review of English literature. Public Admin Policy Rev 11(2):148–168

Mohammadi S, Boni G, Pirlone F, Cattari S, De Angeli S (2023) Multi scale urban modelling to improve the recoverability of cities exposed to multiple natural hazards. In: Compdyn proceedings-9th international conference on computational methods in structural dynamics and earthquake engineering methods in structural dynamics and earthquake engineering, pp 2752–2761

Mulyawati MD, Ferdinand A, Wijaya AK, Saputra HA (2024) Management Planning Water Resources in Balikpapan City, Indonesia: A Case Study. Government & Resilience 2(2):41–62

Nichi J, Cortese TTP (2024) Urban living labs as public policy for smart cities: evidence from tech parks in. Int J Dev Res 14(02):64918–64925

Nowak MJ, Monteiro R, Olcina-Cantos J, Vagiona DG (2023) Spatial planning response to the challenges of climate change adaptation: an analysis of selected instruments and good practices in Europe. Sustainability 15(13):10431

Oikonomaki E, Papadaki I, Kakderi C (2024) Promoting green transformations through smart engagement: an assessment of 100 citizen-led urban greening projects. Land 13(4):556

Ojo B (2024) Strategies for the optimization of critical infrastructure projects to enhance urban resilience to climate change. J Sci Eng Res 11:107–123

Pascariu GC, Banica A, Nijkamp P (2023) A meta-overview and bibliometric analysis of resilience in spatial planning–the relevance of place-based approaches. Appl Spat Anal Policy 16(3):1097–1127

Peterson GD, Cumming GS, Carpenter SR (2003) Scenario planning: a tool for conservation in an uncertain world. Conserv Biol 17(2):358–366

Prashar N, Lakra HS, Shaw R, Kaur H (2023) Urban flood resilience: a comprehensive review of assessment methods, tools, and techniques to manage disaster. Progress Disaster Sci 20:100299

Pratomo RA, Islamiah Z, Bhaskara BE (2024) A pathway to urban resilience: investigating green open space priority areas to mitigate the urban heat island phenomenon in Samarinda City, Indonesia. Int J Disaster Resil Built Env

Qi YD, Xiao X (2020) Transformation of enterprise management in the era of digital economy. Manage World 36(6):135–152

Ragazou K, Zournatzidou G, Sklavos G, Sariannidis N (2024) Integration of circular economy and urban metabolism for a resilient waste-based sustainable urban environment. Urban Sci 8(4):175

Ratican J, Hutson J, Wright A (2023) A proposed meta-reality immersive development pipeline: generative ai models and extended reality (XR) content for the metaverse. J Intell Learn Syst Appl 15:24–35

Razavi H, Titidezh O, Asgary A, Bonakdari H (2024) Building resilient smart cities: the role of digital twins and generative AI in disaster management strategy. In: Digital twin computing for urban intelligence. Singapore, Springer Nature Singapore, pp 95–118

Reinwald F, Weichselbaumer R, Schindelegger A, Damyanovic D (2024) From strategy to implementation: mainstreaming urban green infrastructure in Austria's spatial planning instruments for climate change adaptation. Urban For Urban Green 94:128232

Rezvani SM, Falcão MJ, Komljenovic D, de Almeida NM (2023) A systematic literature review on urban resilience enabled with asset and disaster risk management approaches and GIS-based decision support tools. Appl Sci 13(4):2223

Rezvani SM, Silva MJF, de Almeida NM (2024) Urban resilience index for critical infrastructure: a scenario-based approach to disaster risk reduction in road networks. Sustainability 16(10):4143

Ribeiro PJG, Gonçalves LAPJ (2019) Urban resilience: a conceptual framework. Sustain Cities Soc 50:1–12

Rosemann M, Becker J, Chasin F (2021) City 5.0. Bus Inf Syst Eng 63:71–77

Sanchez TW, Shumway H, Gordner T, Lim T (2023) The prospects of artificial intelligence in urban planning. Int J Urban Sci 27(2):179–194

Scribano A (2021) City and emotions. In: Scribano A, Camarena Luhrs M, Cervio A (eds) Cities, capitalism and the politics of sensibilities. Cham, Switzerland, Palgrave Macmillan

Selim A, Yousef P, Hagag M (2018) Smart infrastructure by (PPPs) within the concept of smart cities to achieve sustainable development. Int J Crit Infrastruct 14:182–198

Shafik W (2024) Shaping the next generation smart city ecosystem: an investigation on the requirements, applications, architecture, security and privacy, and open research questions. In: Smart cities: innovations, challenges and future perspectives. Cham, Springer Nature Switzerland, pp 3–52

Shahat-Osman A, Elragal A (2021) Smart cities and big data analytics: a data-driven decision-making use case. Smart Cities 4:286–313

Shan JD, Tian J, Zeng J (2022) Planning methods of resilient city to cope with extreme climatic disasters. City Disaster Reduct 5:6–12

Shao Y, Sun Y, Zheng Z (2023) How do comprehensive territorial plans frame resilience? A content analysis of plans by major cities in China. Sustainability 15(10):7783

Shen X, Wang G (2018) "Smart nation" in Singapore: perspective of the whole of government. J Intelligence 37(11):69–75

Shen J, Zhang KY (2020) An empirical analysis of factors leading to typical urban problems in China. Prog Geogr 39(1):1–12

Singh S, Sharma P, Yoon B, Shojafar M, Cho G, Ra I (2020) Convergence of blockchain and artificial intelligence in IoT network for the sustainable smart city. Sustain Cities Soc 63:102364

Sivarethinamohan R, Reddy RS (2024) Digital twin for smart city resilience and solutions. In: digital twin and blockchain for smart cities, pp 605–619

Stojanovic M, Mitkovic P, Mitkovic M (2014) The scenario method in urban planning. Facta Universitatis—series. Arch Civ Eng 12(1):81–95

Suárez M, Benayas J, Justel A, Sisto R, Montes C, Sanz-Casado E (2024) A holistic index-based framework to assess urban resilience: application to the Madrid Region, Spain. Ecol Indic 166:112293

Sun L, Makridis MA, Genser A, Axenie C, Grossi M, Kouvelas A (2024) Antifragile perimeter control: anticipating and gaining from disruptions with reinforcement learning. arXiv preprint arXiv:2402.12665

Svítek M, Skobelev P, Kozhevnikov S (2019) Smart city 5.0 as an urban ecosystem of Smart services. In: Service oriented, holonic and multi-agent manufacturing systems for industry of the future, proceedings of SOHOMA, Shenzhen, China, 15–17 November 2019. Springer, Singapore, pp 426–438

Tan L, Cui Q, Chen L, Wang L (2024) An exploratory study on spatial governance toward urban–rural integration: theoretical analysis with case demonstration. Land 13(12):2035

Todella E, Abastante F, Cotella G (2024) Practicing multilevel governance: the revision of the piedmont regional territorial plan. Land 13(6):755

Tsoutsa P, Panagiotakopoulos T, Damasiotis V, Fitsilis P (2024) Empowering urban futures: key competencies for smart city resilience officers. Urban Governance 4(3):180–192

Varzeshi S, Fien J, Irajifar L (2024) Integrating technology and urban resilience: a comprehensive analysis of smart city initiatives in sydney. Sustainability (2071–1050) 16(24)

Viale Pereira G, Klausner LD, Temple L, Delissen T, Lampoltshammer T, Priebe T (2023) Smart cities and digital twins in lower Austria. In: Proceedings of the 24th annual international conference on digital government research, pp 644–645

Vitale C, Meijerink S, Moccia FD (2023) Urban flood resilience, a multi-level institutional analysis of planning practices in the Metropolitan City of Naples. J Environ Plan Manag 66(4):813–835

Wang HW, Peng ZR, Wang D, Meng Y, Wu T, Sun W, Lu QC (2020) Evaluation and prediction of transportation resilience under extreme weather events: a diffusion graph convolutional approach. Transp Res C Emerging Technol 115:102619

Wang W, Zhu J, Fei Z, Ma D, Du X (2023a) Overall framework and development path of national land space resilience planning and construction. Strategic Study CAE 25(3):209–218

Wang K, Jiang GX, Luo Y et al (2023b) Overview of territorial space planning response to climate change. Planner 39(2):5–10

Wang Y, Li S, Gu X, Sun Q, Zhang R, Chen F (2024) Spatio-temporal evolution and multi-subject influencing factors of urban green development efficiency in China: under the carbon neutral vision constraint. Ecol Indic 167:112580

Wardekker JA, de Jong A, Knoop JM, van der Sluijs JP (2010) Operationalizing a resilience approach to adapting an urban delta to uncertain climate changes. Technol Forecast Social Change 77(6):987–998

Wasserman D, Flaxman M (2022) Artificial intelligence and planning practice. American Planning Association, Chicago

Watson JR, Eisenberg DA, Anderson R, Bhatia U, Chatterjee S, Gonzalez MC et al (2024) Designing, operating, and rebuilding infrastructures and lifelines for resilience to extreme flooding. Front Water 6:1438086

Wilkinson A, Kupers R, Mangalagiu D (2013) How plausibility-based scenario practices are grappling with complexity to appreciate and address 21st century challenges. Technol Forecast Social Change 80(4):699–710

Williams J (2023) Circular cities: planning for circular development in European cities. Eur Plan Stud 31(1):14–35

Wilson J, Daugherty P (2018) Collaborative intelligence: humans and ai are joining forces. https://hbr.org/2018/07/collaborative-intelligence-humans-and-ai-are-joining-forces. Accessed 15 Nov 2023

Wolniak R, Stecuła K (2024) Artificial intelligence in smart cities—applications, barriers, and future directions: a review. Smart Cities 7(3):1346–1389

Wolniak R, Gajdzik B, Grebski W (2023) The implementation of Industry 4.0 concept in smart city. Sci Pap Silesian Univ Technol Organ Manag Ser 178:753–770

Xiong K, Sharifi A, He BJ (2022a) Resilient-smart cities: theoretical insights. In: Sharifi A, Salehi P (eds) Resilient smart cities. The urban book series. Springer, Cham

Xiong J, Lin H, Huang P et al (2022b) Key problems and ideas of formulating technical standards in territorial spatial planning. Urban Plan Forum 6:80–87

Xu YY, Li G, Cui SH et al (2018) Review and perspective on resilience science: from ecological theory to urban practice. Acta Ecol Sin 38(15):5297–5304

Xu XS, Yan Y, Chen XH et al (2023) Framework system and path of smart resilient city construction. Strategic Study CAE 25(1):10–19

Xu H, Omitaomu F, Sabri S, Zlatanova S, Li X, Song Y (2024) Leveraging generative AI for urban digital twins: a scoping review on the autonomous generation of urban data, scenarios, designs, and 3D city models for smart city advancement. Urban Informatics 3(1):29

Xue F, Luan B, Fan Y, Xie S, Yang X, Luo J, Zheng R (2024) Assessing the lifecycle environmental resilience of urban green infrastructures coping with acute disturbances and chronic stresses. Water 16(8):1162

Xuesong X, Yue Y, Xiaohong C, Xingbao L, Yun S, Jiale T, Jianjun P (2022) Framework system and path of smart resilient city construction. Strateg Study Chin Acad Eng 25(1):10–19. https://doi.org/10.15302/J-SSCAE-2022.05.027

Yang G, Zhang P, Yu F, Zhu X (2023) A review on resilient cities research from the perspective of territorial spatial planning: a bibliometric analysis. Front Ecol Evol 11:1300764

Yang W, Xiong Z, Quek TQ, Shen X (2024) Streamlined transmission: a semantic-aware xr deployment framework enhanced by generative AI. IEEE Network

Ye X, Du J, Ye Y (2022) MasterplanGAN: facilitating the smart rendering of urban master plans via generative adversarial networks. Environ Plan B Urban Anal City Sci 49(3):794–814

Yigitcanlar T, Guaralda M, Taboada M, Pancholi S (2018) Place making for knowledge generation and innovation: planning and branding Brisbane's knowledge community precincts. In: Urban knowledge and innovation spaces. Routledge, pp 115–146

Yildirim Y, Keshavarzihaghighi G, Aman AR (2021) Sustainable responses of an urban park for disaster resilience: a case study of hurricane Harvey. Int J Sustain Dev World Ecol 28(8):720–732

Yun Y, Lee M (2019) Smart city 4.0 from the perspective of open innovation. J Open Innov Technol Mark Complex 5:92

Zhang J, Guo X, He W (2024a) Resilience policies in China: an analysis of central-level policies. Policy Stud:1–21

Zhang B, Ma D, Wang W (2024b) Implementing A resistance-relief approach into evaluating urban disaster management capacity: a case study of Xuzhou. Int J Disaster Risk Reduct 103:104348

Zhao RD, Fang CL, Liu HM (2020) Progress and prospects of urban resilience research. Prog Geogr 39(10):1717–1731

Zheng Y, Lin Y, Zhao L, Wu T, Jin D, Li Y (2023) Spatial planning of urban communities via deep reinforcement learning. Nat Comput Sci 3(9):748–762

Zhou K, Wang S, Feng Y (2023) How is spatial planning adapting to climate change? A textual analysis based on the territorial and spatial plans of 368 Chinese cities. Land 12(11):1993

Zhuang Y, Cenci J, Zhang J (2024) Review of big data implementation and expectations in smart cities. Buildings 14(12):3717

Zubizarreta I, Seravalli A, Arrizabalaga S (2016) Smart city concept: what it is and what it should be. J Urban Plan Dev 142:04015005

Interdisciplinary Studies on Urban Resilience: Urban Digital Twins and Construction 5.0 for Resilient, Sustainable and Smart Cities

9

Abstract

The SDuMuRS framework emphasizes sustainability, durability, multi-hazard resistance, resilience, and smart monitoring guides the design of future urban infrastructures. This approach aligns with global sustainability objectives, including net-zero emissions by 2050 and the United Nations' Sustainable Development Goals (SDGs). Urban resilience, defined through engineering, ecological, and socio-ecological perspectives, focuses on a city's capacity to withstand and adapt to various shocks. Urban Digital Twins (UDTs) emerge as pivotal tools in smart city development. These digital replicas of urban environments facilitate real-time interaction and decision-making, enhancing infrastructure management and sustainability efforts. UDTs integrate data from various sources, enabling scenario modeling, emergency preparedness, and infrastructure optimization. However, challenges such as data management, interoperability, cybersecurity, and public engagement persist. Construction 5.0 represents the next evolution in the architecture, engineering, and construction (AEC) sector, emphasizing human-centric technology, resilience, and sustainability. It integrates digital technologies like AI, IoT, and blockchain to enhance efficiency and reduce environmental impacts. Key aspects include adaptive technologies for risk mitigation, energy-efficient designs, smart buildings with automated systems, circular economy principles, and automation in construction workflows. Despite its potential, challenges such as high initial investment costs, workforce upskilling, data security concerns, and regulatory constraints need to be addressed. Integrating resilience into urban planning is crucial, especially in rapidly urbanizing areas with weak institutional frameworks and vulnerability to climate change. Failure to provide essential services exacerbates social and economic vulnerabilities. Aligning urban development with the SDGs, particularly SDG 11, advocates for inclusive, clean, resilient, and sustainable cities. Urban digital twins and Construction 5.0 offer frameworks for integrating resilience and sustainability into urban planning and construction practices.Future research should explore human-centered approaches and interdisciplinary methodologies to optimize urban planning and construction practices, ensuring long-term sustainability and resilience in urban environments.

9.1 Introduction

The design and construction of future civil engineering structures should prioritize sustainability, durability, multi-hazard resistance, resilience,

V. Yorucu et al., *Modern Approach of Resilient and Sustainable Smart Cities*, Sustainable Development Goals Series, https://doi.org/10.1007/978-3-032-05715-0_9

and smart monitoring and management—collectively known as the SDuMuRS framework, as proposed by Hao et al. (2023).

The design and construction of future civil engineering structures must prioritize sustainability, durability, multi-hazard resistance, resilience, and smart monitoring and management, referred to as SDuMuRS design. As the world rapidly changes, there is an increasing demand for a sustainable and resilient built environment. This global shift is supported by numerous priorities and initiatives set forth by organizations like the United Nations and various governments, such as the goal of achieving net-zero emissions by 2050 and the Sustainable Development Goals (SDGs). As a result, our infrastructure and structures need to be sustainable and durable to mitigate pollution, global warming, and the depletion of limited natural resources. Besides resisting various natural and man-made hazards, our infrastructure must also adapt to the changing global climate and other evolving conditions (Hao et al. 2023).

Designing sustainable cities for the future involves integrating lessons from the past and anticipating future challenges by articulating a vision that encompasses multiple scales and key principles such as energy, ecology, infrastructure, waste, livability, mobility, accessibility, economy, and culture. This must be done while addressing major societal and intellectual trends and paradigm shifts in science and technology. Sustainable cities have become a dominant global paradigm in urbanism, with eco-cities being one of the most advocated models for sustainable urban development (Bibri 2023). It is argued that the eco-city model can meet key environmental sustainability objectives and provide economic and social benefits (Bibri and Krogstie 2021). However, discussions around eco-cities often emphasize environmental goals more than economic and social goals (Bibri 2021a, b).

Eco-city projects are gaining prominence in strategic policy, enviro-economic, and politico-economic discourses worldwide, becoming a significant focus in various research fields. It is crucial to use modern ecological civilization and ecological restoration as guides for managing all economic and social activities in cities. Cities are increasingly central to efforts to transition toward a sustainable, low-carbon society, with many city governments and municipal agencies promoting eco-city or eco-district initiatives (Bibri 2023).

The United Nations' 2030 Agenda highlights the role of advanced ICT in promoting socio-economic development, protecting the environment, increasing resource efficiency, upgrading infrastructure, and retrofitting industries based on sustainable design principles. This aligns with the smart city's potential to use big data technologies for strategic sustainable development. The primary goal of smart cities is to enhance economic development, quality of life, and various sustainability targets through data and technology (Ahvenniemi et al. 2017). The rapid growth of urban data and its analytical capabilities creates new opportunities for innovation in eco-cities, leading to more effective ways to integrate sustainability into environmental, economic, and social aspects. Bibri (2023) created an integrated model for strategic sustainable urban development, outlining a series of actions and measures necessary to transform the green, sustainable, smart, economic, and social infrastructures of data-driven smart eco-cities of the future.

Resilient cities are those that can absorb, recover from, and prepare for future economic, environmental, social, and institutional shocks, while also promoting sustainable development, well-being, and inclusive growth. The literature continues to explore the concepts of urban resilience, resilient cities, smart cities, and smartness, with ongoing discussions and varying interpretations. In the modern approach to economic resilience, smart city resilience involves achieving and surpassing economic, social, and cultural performances to establish a new developmental balance. This balance is founded on combining various factors and resources into a new growth pattern, leveraging past experiences—both positive and negative—and reconstructing subsystems through creative disruption. This leads to a robust and sustainable recovery and smart development (Apostu et al. 2022).

Tzioutziou and Xenidis (2021) offer a recent synthesis of the literature, investigating the potential convergence between resilience and smart city frameworks. They identify common components and specific characteristics of each, noting that both concepts aim for sustainability and share a systems-thinking approach. They also point out that adaptation, efficiency, and knowledge creation are integral to both urban resilience and smartness. Resilience is increasingly viewed as an appealing perspective for cities, conceptualized as highly complex, adaptive systems. Smart city resilience, especially in the context of economic resilience, involves achieving and surpassing economic, social, and cultural performance through a new developmental balance. This balance is based on leveraging past experiences and rebuilding subsystems for robust and sustainable recovery and smart development. It represents a new developmental paradigm where economic resilience is fully integrated as an objective of human development, not just mechanical resilience.

The evolution of urban models is well-documented in historical reviews, such as the one by Boyce and Williams (2015) on urban transportation modeling, and a broader review by Batty (2008) on urban models. These models cover a range of topics, including transportation, land use and urban growth, and economic development. Traditionally, urban infrastructure models were developed separately because they had different backgrounds and used different methods. Technically oriented urban models have encompassed systems like road infrastructure, water supply, and electric power transmission and distribution. In some instances, 3D city models were also created. Although digital models were adopted early, it wasn't until the 2000s that city plans began to be replaced by 3D city models. These models were further refined in the 2010s with the incorporation of Building Information Modelling (BIM) data. During this period, the concept of the smart city—where city administration and citizens use new technologies to make the city more efficient, intelligent, sustainable, safer, inclusive, and democratic—gained popularity, and cities began to use sensors.

Urban digital twins (UDT) have the potential to advance the smart city concept and urban models to a new level. They leverage data gathered in smart cities, automatically integrating it into the city model and its systems to maintain an accurate digital replica that can autonomously interact with the city (Ferré-Bigorra et al. 2022). City digital twins can model, mirror, and interact with not only the physical aspects of the city but also its social and economic dimensions (Wan et al. 2019). The continuous and bidirectional data exchange enables the digital twin to be the most accurate representation possible of the real city and its systems. This interactivity offers significant benefits for urban infrastructure management and operations. Although several digital twins of cities have been developed, each implementation has been unique, and there is no consensus on what an UDT twin should encompass or which systems it should model (Ferré-Bigorra et al. 2022).

A clear understanding of urban resilience, sustainability and smart cities is essential to the development of DT in urban planning. Additionally, incorporating social community dynamics into UDT is an emerging area of research in the context of resilient and sustainable smart cities. This study presents insights with some knowledge gaps in understanding of the Urban Resilience vs. Smart City Resilience—Discussions and Advancements, Urban Digital Twins for Resilient, Sustainable and Smart Cities, and Construction 5.0 for Resilient, Sustainable and Smart Cities.

9.2 Urban Resilience Vs. Smart City Resilience

Urban resilience is the ability of an urban system, including socio-ecological and socio-technical networks, to maintain or quickly return to its initial functions and to transform systems for current or future adaptation. Urban resilience is based on four pillars: recovery, adaptation, transformation, and resilience. It can be defined from different perspectives:

- Engineering perspective: the capacity of a city to absorb change or stress and return to its previous state.

- Ecological perspective: the capacity of a city to adjust to a shock or disaster without damaging existing structures and relationships.
- Socio-ecological (evolutionary) perspective: the capacity of a city to adapt or transform in response to change or shock.

De Falco (2018) note that smart cities and resilient cities share similar development trajectories but are interconnected. Some smart cities, however, may not be more resistant to unexpected events like climate change or disasters.

Creating a smart city is a strategy to address issues arising from urban population growth and rapid urbanization. Chourabi et al. (2012) identifies eight critical factors for smart city initiatives: management and organization, technology, governance, political context, people and communities, economy, infrastructure, and the natural environment. These factors form an integrative framework for examining local government initiatives regarding smart cities and have practical implications for government professionals.

Urban areas are influenced by a range of factors, including rapid urbanization, urban regeneration, immigration, and economic cycles (LopezDeAsiain and Díaz-García 2020). The complexity of urban systems is further compounded by natural hazards. This is particularly critical in cities within emerging economies experiencing rapid urbanization, often marked by inadequate planning, weak institutional frameworks, and a lack of essential public services (Sarker et al. 2020a, b). Due to limited capacity, political will, or funding to address climate change, preventive actions are only considered significant or feasible if they result in improved urban quality of life. A failure to provide basic services exacerbates social and economic vulnerability in urban areas, while natural hazards further increase vulnerability and reduce resilience (Sarker et al. 2021).

Resilience and sustainability are viewed as effective strategies for addressing hazards and enhancing urban planning processes (Pirlone et al. 2020). Sustainable Development Goals (SDGs) emphasize the connection between sustainability and resilience, and scholars recognize the importance of understanding these concepts for related fields (Fitzgibbons and Mitchell 2019). Redman (2014) noted that sustainability and resilience are complementary, occasionally used interchangeably, and share several principles. Zhang and Li (2018) pointed out that both focus on a city's vulnerability and capacity to endure, with overlapping meanings, though this overlap can sometimes weaken their distinctiveness. There are varied definitions and applications of resilience and sustainability in literature. Miller (2013) explored alternative approaches to sustainability, while Brand and Jax (2007) argued that ambiguous and unclear interpretations of these concepts hinder scientific progress due to their complexity.

Resilience refers to a system's ability to "bounce back" or return to a stable state after experiencing stressors from hazards (Sarker et al. 2020a, b). It also involves the capacity of urban communities to recover from risks and adapt to changes in response to internal and external hazards (McGill 2020). Developing resilience thinking is crucial for addressing environmental risks amidst global environmental changes. Numerous cities have undertaken plans, programs, and initiatives to integrate sustainability into urban planning (Monstadt and Schmidt 2019), and resilience planning is advancing in both developed and developing countries.

The SDGs specifically highlight the need for cities to be "inclusive, clean, resilient, and sustainable" (SDG 11). Urban management plays a crucial role in global efforts to mitigate disaster risks and the adverse effects of climate change. New urban policies should emphasize resilience, reflecting their importance as a conceptual framework for envisioning the future of cities (da Silva et al. 2019).

Understanding urban resilience requires familiarity with the evolution of resilience theory (Meerow and Stults 2016). Although the concept has been used in psychology, medicine, and engineering for some time, it is often attributed to ecologist C.S. Holling in the context of global environmental change. Holling (1973) described resilience as an ecosystem's capacity to maintain essential functions despite disturbances. The socio-ecological system (SES) theory, developed by interdisciplinary ecologists, extended Holling's ecological principles to the social realm, viewing nature and society as interconnected and coevolving structures. Resilience is also seen as a system's ability to endure perturbations without losing key functions or changing states, its potential to self-organize, and its capacity for adaptation (Meerow et al. 2016).

9.3 Urban Digital Twins for Resilient, Sustainable and Smart Cities

The beginning of the twenty-first century has seen a remarkable increase in urbanization, presenting both challenges and opportunities for cities around the globe (Major et al. 2021; Mohammadi and Taylor 2019). As noted by Deng et al. (2021), this growth in urban areas involves a complex mix of socioeconomic, technological, and environmental factors rather than being merely a demographic shift. As cities expand, it becomes crucial to manage and support this growth while ensuring sustainability, resilience, and a high quality of life for residents (Banihashemi and Sohi 2022; Mohammadi and Taylor 2019).

The concept of 'smart cities' has gained traction, focusing on the use of digital tools and data-driven strategies to improve urban living (Mylonas et al. 2021). In this digital context, Digital Twins (DTs) are increasingly recognized and applied in urban planning and infrastructure management (Depretre et al. 2022; Deren et al. 2021; Elsehrawy et al. 2021). As explained by Elsehrawy et al. (2021), DTs are virtual models of real-world entities, creating a connection between the digital and physical aspects of urban environments. Although still emerging, this technology shows great potential in urban planning by offering interactive, real-time visualizations and simulations of urban settings (Caprari et al. 2022). DTs provide a comprehensive, data-driven approach to tackling urban issues, ranging from infrastructure planning to environmental sustainability (Caprari et al. 2022).

The urban management sector has recently shown a growing interest in utilizing DT to enhance urban planning, optimize asset management, and create safer, more sustainable cities. Building on the insights gained from developing smart cities and applying DT in other industries, urban digital twins (UDT) have rapidly expanded in recent years. An UDT is a detailed digital representation of urban environments, including their components, operations, and systems (Nochta et al. 2021). It aids in decision-making processes to achieve city-scale outcomes such as urban planning and management, as well as related services, providing better perspectives for making informed decisions (Nochta et al. 2021). An UDT consists of interconnected sub-DTs, each representing specific aspects of the urban environment's functioning and development (Lu et al. 2020; Papyshev and Yarime 2021). These are integrated with advanced features like AI, machine learning, and data analytics, enabling precise adjustments and alignment with the city's actual infrastructure conditions by integrating data from various sources in real time (Ivanov et al. 2020). These models are designed to accurately and quickly reflect and predict the current and future conditions of their physical counterparts (Lu et al. 2020). UDT can elevate the smart city concept and urban models by integrating data collected in smart cities into a comprehensive digital replica that can autonomously interact with the city's systems (Ferré-Bigorra et al. 2022). UDT can not only model and mirror the physical city but also focus on its social and economic aspects (Wan et al. 2019). The continuous, two-way data exchange ensures that the DT accurately represents the real city and its systems, potentially offering significant benefits for managing and operating urban infrastructure.

9.3.1 Development Phases of Urban Digital Twins

Developing a UDT involves a complex integration of physical and virtual realms. Petrova-Antonova and Ilieva (2019) outline a six-phase process for this development: creation, interaction, acquisition, analysis, comprehension, and decision. In the 'Creation' phase, a City Information Model (CIM) is assembled using data from various sources, such as sensor readings and existing municipal information systems. The 'Interaction' phase establishes real-time, bidirectional communication between the city's physical and virtual counterparts, utilizing advanced communication and edge computing technologies. During the 'Acquisition' phase, the collected data is centralized and refined for analysis. In the 'Analysis' phase, AI and cognitive computing techniques are used to extract meaningful insights. The 'Comprehension' phase then visualizes these insights through formats like 3D models and dashboards, highlighting potential areas for improvement. Finally, the 'Decision' phase turns these insights into actionable strategies, aiming to enhance the operations of the physical city to align with the efficiency and optimization demonstrated in its DT (Petrova-Antonova and Ilieva 2019). The phases of UDT development for smart cities are depicted in Fig. 9.1.

9.3.2 Application for Urban Digital Twins in Smart City Realm

UDT serves as a practical bridge between the theoretical concepts of smart cities and actual interventions, optimizing the performance of a smart city (Petrova-Antonova and Ilieva 2019). They can capture real-time data (Shirowzhan et al. 2020), providing a comprehensive visualization of all city resources and interactions. This capability aids in monitoring infrastructure, utilities, businesses, and planning future developments (Mashaly 2021). UDTs offer insights into urban efficiency, leading to real-world actions such as changes in city design or transportation methods (Petrova-Antonova and Ilieva 2019). Mylonas et al. (2021) discuss various smart city domains where UDTs can be applied, including building surveillance, urban planning, circular economy, transport, risk mitigation, healthcare, and sustainability. Relatedly, Alva et al. (2022) outline six use cases for UDTs, such as predictions and scenario modeling, emergency preparedness, optimizing functionalities, and strategic planning. Further studies by Yang and Kim (2021), Wang and Vu (2023), Ferré-Bigorra et al. (2022), and Elsehrawy et al. (2021) explore UDT applications in urban development, administration, transport, environment, energy, and disaster management.

The development and implementation of UDT for smart city governance and urban planning face numerous challenges, crucial for assessing their sustainability and practicality. According to Lei et al. (2023), these challenges include data management and integration, data quality and synchronization, interoperability, standards and connectivity, cybersecurity and data privacy, data modeling, utilization and visualization, technical complexity and infrastructure, financial constraints, resource allocation, and socio-political implications and public engagement as listed Table 9.1.

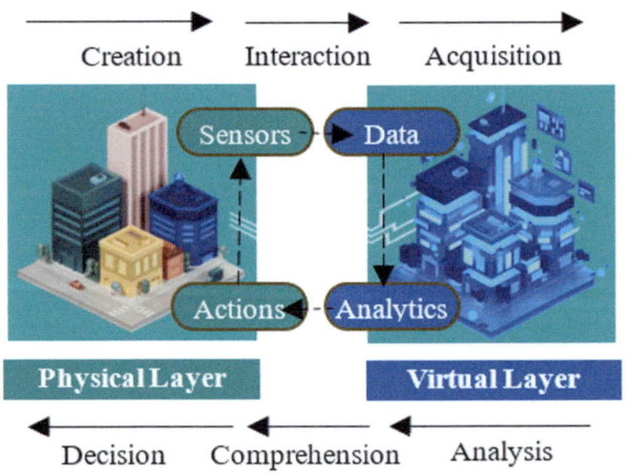

Fig. 9.1 Phases of UDT development for smart cities. (Source: Author's compilation)

Table 9.1 Challenges in potential applications of UDT for smart city governance and urban planning

Challenge	Description	References
Data management and integration	City data is complex, requiring powerful computing and standardized models Data integration in DT development is challenging due to source heterogeneity	Peldon et al. (2024), Shahat et al. (2021), and Lu et al. (2020)
Data quality and synchronization	Ensuring data quality is crucial for effective DT utility and integrity Real-time data synchronization challenges can compromise speed and quality	Peldon et al. (2024), Weil et al. (2023), and Lu et al. (2020)
Interoperability, standards and connectivity	Lack of standards hinders UDTs' data exchange Limited interoperability restricts effective UDT data exchange Data misinterpretation may cause safety risks and suboptimal city performance Interoperability challenges affecting data connectivity hinder data exchange between smart cities	Peldon et al. (2024), Weil et al. (2023) Ferré-Bigorra et al. (2022), Petrova-Antonova and Ilieva (2019), and Wang and Vu (2023)
Cybersecurity and data privacy	Cybersecurity is crucial, as it breaches risk data privacy and control over infrastructure Ambiguous data ownership and privacy concerns complicate data usage understanding	Peldon et al. (2024), Ferré-Bigorra et al. (2022), and Petrova-Antonova and Ilieva (2019)
Data modelling, utilization and visualization	Constrained to static data management, BIM and GIS need additional tools to manage real-time data. Current research integrating BIM, GIS, AR, and VR in unstructured environments overlooks data privacy concerns UDT face issues with model accuracy and data quality, affecting visualization Determining needed data types is challenging due to insufficient computing resources	Peldon et al. (2024), Wang and Vu (2023), Shahat et al. (2021), and Wang and Vu (2023)
Technical complexity and infrastructure	UDT requires a skilled, sizable workforce for design and maintenance UDT infrastructure includes hardware, software, and networking for real-time data processing	Peldon et al. (2024), Ferré-Bigorra et al. (2022), and Weil et al. (2023)
Financial constraints and resource allocation	Limited budgets for UDT require balancing cost and functionality	Peldon et al. (2024) and Ferré-Bigorra et al. (2022)
Socio-Political implications and public engagement	Public and departmental involvement faces challenges from missing governance and data-sharing frameworks Socio-political impacts of UDT need analysis to identify acceptance challenges	Shahat et al. (2021) and Ferré-Bigorra et al. (2022)

Source: Author's compilation

9.3.3 Proposed UDT Structure for Smart and Resilient Cities

The structure of UDT is largely determined by their intended use, including the users, life-cycle stage, and specific applications, making it challenging to describe a standard digital twin framework. Despite these differences, all city DT share a common feature: a digital model of the physical city. This model serves as the foundational element for building the urban digital twin. Interactive capabilities are introduced by adding sensors and actuators to the model, transforming it into a digital twin. A fully developed UDT is achieved by incorporating various other infrastructures, services, systems, and their corresponding models, data sources, and actuators (Ferré-Bigorra et al. 2022; Peldon et al. 2024).

The UDT city model, along with the proposed modules and individual systems, is organized into four functional layers: the data acquisition layer, data modeling layer, simulation layer, and service/actuation layer. Additionally, there is a physical layer that consists of the physical systems or services represented by the digital twin module. The components within each layer vary depending on the physical systems depicted in a module. To accurately reflect the physical layer, the data acquisition layer is responsible for automatically capturing and transmitting data to the digital modeling layer, where the digital replica of the real system is continuously updated. The simulation layer then processes the data within the model and sends the results to the service/actuation layer. Here, the digital twin interacts with the real system either through direct actuation or indirectly by providing data to users (Ferré-Bigorra et al. 2022; Peldon et al. 2024).

UDT, when properly implemented, can provide significant benefits to various stakeholders in a city, including citizens, public administration, asset managers, asset owners, and researchers. For public administration, urban digital twins enable more informed decision-making in city planning and operations, leading to better urban management and a more sustainable city. This improved management positively impacts the whole society. Additionally, DT can simultaneously model multiple systems and process data from various datasets, allowing for integrated city planning and control within a single tool. This unification simplifies data use for both basic users, such as citizens or tourists, and expert users, such as asset managers or researchers. Making data more accessible to citizens also encourages greater participation and engagement in urban planning. A unified city digital twin is a more efficient urban management tool than traditional methods, as it can be regularly updated to reflect the city's ongoing evolution, providing a sustainable alternative to constantly updating separate models. Moreover, this unified approach allows for easier identification of relationships and trends between systems, enhancing data utilization. Finally, UDT serves as strategic planning platforms where public and private initiatives can be tested and coordinated (Ferré-Bigorra et al. 2022; Peldon et al. 2024). The proposed UDT structure for smart and resilient cities is illustrated in Fig. 9.2.

9.3.4 Real-World Use Cases of Urban Digital Twins in Smart Cities

Urban Digital Twins (UDTs) are emerging as powerful tools for translating the theoretical framework of smart cities into actionable interventions that improve city operations and resilience. In Singapore, the "Virtual Singapore" initiative exemplifies how UDTs can support real-time data integration and 3D visualization for urban planning. This dynamic platform enables government agencies, businesses, and researchers to simulate energy use, pedestrian flows, and infrastructure impacts, allowing for evidence-based planning and efficient resource allocation. Similarly, in Hangzhou, China, the City Brain platform—developed by Alibaba—uses real-time traffic data from cameras and sensors to dynamically adjust traffic lights and prioritize emergency response routes, demonstrating how UDTs can optimize transportation and urban mobility.

In the domain of risk mitigation, Helsinki has adopted a UDT framework to model flood scenarios and heat stress, using this data to guide climate adaptation strategies. This application reflects the value of digital twins in environmental resilience and emergency preparedness. Rotterdam, in the Netherlands, offers another compelling use case by integrating a UDT to simulate the effects of new building developments on traffic, energy consumption, sunlight exposure, and noise levels. These simulations inform zoning and land-use decisions, supporting sustainable urban development. In the healthcare sector, New York City employs digital infrastructure to monitor hospital capacity and coordinate ambu-

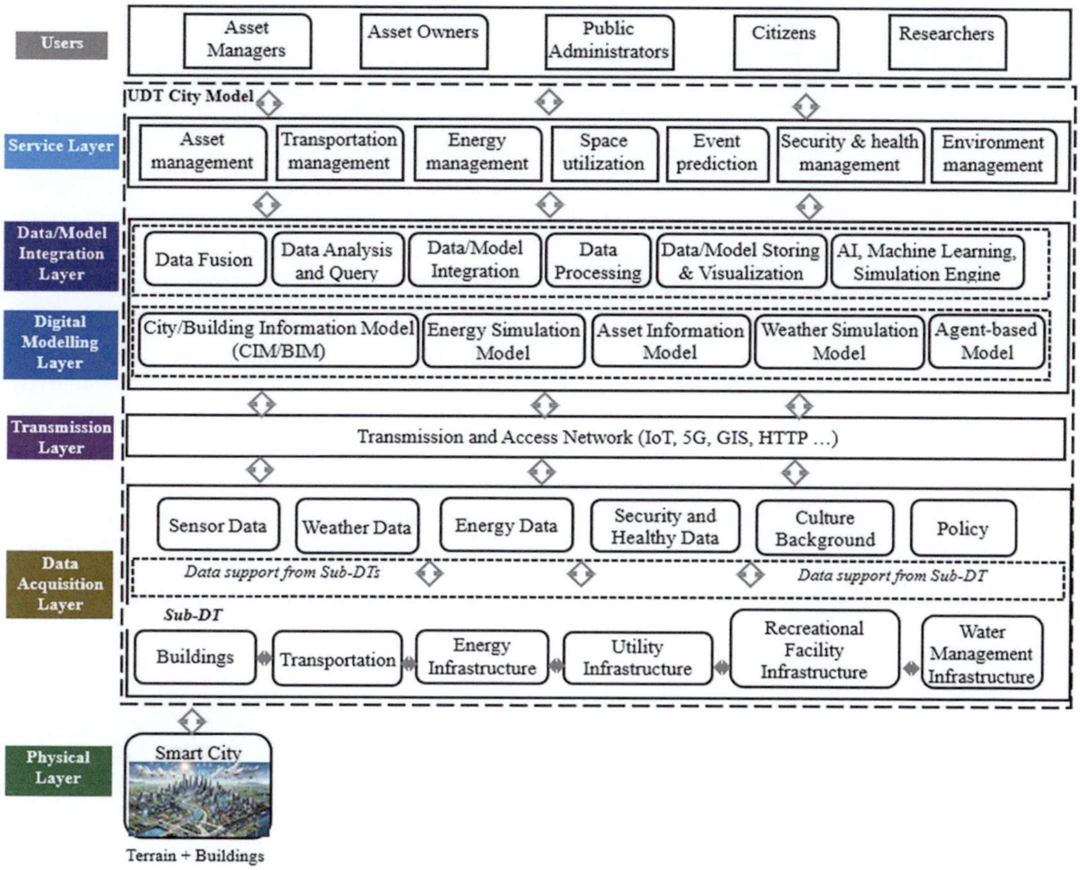

Fig. 9.2 Proposed UDT structure for smart and resilient cities. (Source: Author's compilation)

lance routing, ensuring a timely response to medical emergencies and optimizing resource deployment.

UDTs also play a critical role in advancing the circular economy. In Glasgow, a digital twin model is used to track construction materials and emissions, supporting the reuse of building materials and aligning with sustainability goals. Governance applications are equally prominent, as seen in Barcelona, where UDTs are embedded in the city's smart governance framework. Citizen feedback, e-governance platforms, and real-time sensor data are integrated to inform public service delivery and planning decisions.

Emerging regions are also exploring the potential of UDTs. In Northern Cyprus, a pilot initiative is to be developed in the Catalköy–Esentepe region. This project aims to use digital twins to simulate seasonal tourism impacts on energy grids, water networks, and coastal infrastructure. The insights gained would support spatial planning, resource optimization, and climate resilience strategies.

These diverse use cases reflect the adaptability of UDTs across sectors including mobility, planning, climate resilience, public health, governance, and sustainability as stated earlier by Yitmen and Alizadehsalehi (2021). However, the successful deployment of UDTs depends on addressing key challenges such as data integration, interoperability, infrastructure, cybersecurity, and public engagement—as outlined in Table 9.1. Collectively, these examples demonstrate how UDTs serve as a practical bridge between digital innovation and smart city governance, enabling cities to become more responsive, resilient, and efficient.

9.4 Construction 5.0 for Resilient, Sustainable and Smart Cities

Ikudayisi et al. (2023) suggest that future research on Construction 5.0 should focus on human-centric technology, resilience, and sustainability in the Architecture, Engineering, and Construction (AEC) industry. Yitmen et al. (2023) developed a model and system for Construction 5.0, marking the first study to explore the relationships between human-centric technology, resilience, sustainability, and the opportunities and challenges of implementing Construction 5.0 for smart, sustainable, and resilient buildings using the Structural Equation Modeling (SEM) method.

In the context of Construction 5.0, resilience refers to the industry's capacity to adapt and recover from disruptions. This involves incorporating technologies that enable quick adaptation to changing conditions, robust planning strategies, and systems that anticipate and mitigate risks. It also includes the ability to recover from unexpected setbacks and maintain continuous operations. Given the AEC industry's role in constructing critical infrastructure, achieving resilience by integrating design, construction, and delivery is crucial. Future efforts should focus on developing practical roadmaps and frameworks to enhance stability across various project types and geographic regions. Integrating resilience training into industry practices is essential. One promising approach is the development of an integrated climate vulnerability index and the application of adaptation and mitigation strategies. Successful implementation requires comprehensive data collection, analysis, documentation, regular updates, and practical application throughout all project stages. Potential research avenues include integrating design and planning processes to enhance project resilience, recovery, and reconstruction efforts, exploring multi-sourcing strategies, implementing Modular Integrated Construction (MiC) systems to optimize reconstruction efficiency, using integrated risk contracting, leveraging big data analytics for resilience planning, and utilizing Building Information Modeling (BIM) to enhance resilience planning (Ikudayisi et al. 2023).

Resilience is a key aspect of Construction 5.0, focusing on designing buildings and systems that can withstand natural disasters and adapt to changing environmental conditions. Resilient buildings are designed to minimize damage and quickly recover from disruptions (Yitmen et al. 2023; Gholami et al. 2021; Roostaie and Nawari 2022). Construction 5.0 also promotes incorporating smart building features such as automated systems, IoT-enabled sensors, and data analytics to optimize energy consumption, comfort, and safety, thereby enhancing occupant experience and operational efficiency (Ikudayisi et al. 2023; Yitmen et al. 2023; Almusaed and Yitmen 2023).

Sustainable building practices in Construction 5.0 include green building certifications, energy-efficient design strategies, integrating renewable energy sources, and responsible material procurement. These practices help significantly reduce a building's environmental footprint and contribute to a greener future (Yitmen et al. 2023).

Maintaining resilience in the AEC industry is crucial for advancing Construction 5.0 and achieving smart, sustainable, and resilient structures. This requires reevaluating current strategies to reduce supply chain vulnerabilities and improve risk management. The focus should be on stability and the implementation of automated solutions for significant changes, driven by technological advancements and fostering an Operator 5.0 workforce adept at navigating the digital landscape. Society 5.0 is committed to sustainable practices and circular economy principles. Resilient buildings should feature disaster preparedness, adaptive design, backup systems, and climate resilience to ensure readiness for unforeseen challenges while maintaining structural integrity and community engagement.

Sustaining resilience in the AEC industry shows a commitment to environmental responsibility and promotes Construction 5.0's goal of smart, sustainable, and resilient structures. This vision includes adopting sustainable building techniques, decentralizing links between production resources and products, and enabling large-scale customization of goods and services. Success in construction depends on balancing economic, environmental, and social aspects throughout project phases. Blockchain technology can help communicate social sustainability and material traceability in supply chains. Achieving Construction 5.0 involves enhancing quantity, speed, quality, and cost reduction through sustainable practices, relying on technological advancements and the rise of Operator 5.0. Society 5.0 embraces sustainability, circular economy principles, and sustainable construction methods focusing on energy efficiency, automation, occupant comfort, and data utilization. Sustainable buildings emphasize energy performance, water efficiency, responsible materials, and interior quality, integrating site selection, land use, and accessible transportation as crucial elements (Yitmen et al. 2024).

The AEC industry aims to redefine construction, creating a built environment aligned with community well-being and values. To sustain resilience in the AEC industry and advance Construction 5.0, it is essential to rethink existing methods to minimize supply chain vulnerabilities and enhance the industry's ability to manage uncertainties. This requires a comprehensive approach to building smart, sustainable, and resilient structures, encompassing energy efficiency, automation, occupant comfort, and technology integration. Key challenges include high initial costs, securing funding, developing a skilled workforce, ensuring data security and privacy, establishing regulatory frameworks, managing scalability, and fostering stakeholder acceptance and adaptability. Addressing ethical concerns about the impact of automation and AI on employment and decision-making is also crucial. Resilient buildings must be designed with disaster preparedness, adaptable features, backup systems, and climate resilience, ensuring structural integrity and community engagement (Yitmen et al. 2024).

Transforming the construction sector to embed environmental and social sustainability and resilience within its processes, business models, and daily decision-making ultimately requires changes in human behavior. While perspectives that examine technology, planning processes, policy, and other aspects contribute valuable knowledge, focusing on the individual actor offers additional important insights. However, the roles of individuals and social groups (such as organizations, projects, and communities) are often under-researched in the context of sustainable and resilient construction. This theme invites papers that use theories and methods from the social sciences to investigate the construction sector (Murtagh et al. (2020).

Salas and Yepes (2019) explore this area by developing decision support systems for strategic urban infrastructure planning. They use a decision support framework for strategic decisions on infrastructure, extending it with applications that evaluate alternatives and generate policy measures considering both risks and opportunities. They applied their models to Spanish cities as a case study, offering guidance to policymakers in evaluating risks and opportunities of planning alternatives under extreme scenarios. By incorporating urban vulnerability assessments into urban infrastructure planning, they make resilience a part of planning processes, addressing the principle of integrated planning, which is crucial for resilient cities (Lloyds 2017). The proposed methodology improves integrated infrastructure planning throughout a territory by maximizing adaptation capacity at the local (city) level while minimizing risks at the national level.

In the construction sector, considering the critical role of human relationships is essential for transitioning to a more sustainable and resilient built environment. There is a need for holistic perspectives at various levels, including:

- The building product (e.g., office blocks, schools, single-family homes) from design to construction, operation, demolition, and reuse.
- The supply chain, which includes all actors and flows (materials, information, finance), is necessary to produce a final building product.
- The urban built environment, including strategic infrastructure, planned at complexity levels beyond single construction projects.

Green building construction differs from traditional methods and may involve non-standard materials or systems. Enhancing energy, water,

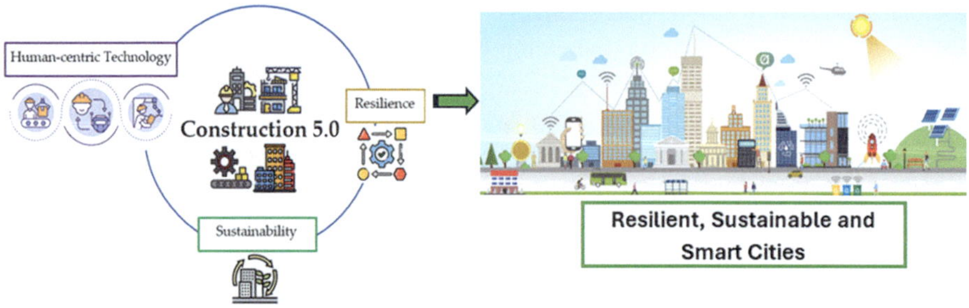

Fig. 9.3 Facilitating construction 5.0 for resilient, sustainable and smart cities. (Source: Author's compilation)

and material efficiency while reducing buildings' adverse impacts on human health and the environment can be achieved through better contextualization of a building's design, construction, operation, maintenance, and end-of-life removal. As holistic and realistic whole-life costs are considered, a compelling business case for green buildings emerges.

For promoting low-carbon construction, supportive policies and incentives from the government are essential. Innovations in technology and materials and advances in design are necessary. Detailed and specialized investigations into specific processes are more effective in reducing emissions in the construction sector than more generic approaches. In the later stages of the production lifecycle, building occupants' behaviors significantly impact, and policies should address this. Achieving low-carbon construction requires a comprehensive understanding of the mutual effects of various stakeholders, including those involved in both production and consumption, and the joint efforts of government and industry.

Understanding of value in sustainable and resilient construction changes over time, which also applies to environmental assessments. As technology, processes, and understanding of environmental sustainability evolve, environmental assessments too. For example, asbestos was once valued for its fire-resistance, thermal insulation, and strength, but is now considered too hazardous to use. Similarly, petrochemical-based insulation, currently efficient, may eventually be seen as too harmful. The point is that cleaner production in the built environment is an ongoing process, and our understanding of sustainability, resilience, environmental impact, and social value, as well as our evaluation methods, must be dynamic rather than static.

For low-carbon construction, the choice of energy sources is a significant factor. In the context of the ongoing energy transition, the use of renewable energy in the construction industry is essential. Renewable energy can be utilized for both material production and operational energy consumption. However, the specific paths and contributions of renewable energy for these purposes need to be investigated and quantified. Additionally, different development stages and resources available in rural and urban areas will influence their optimal energy choices in construction. Identifying targeted energy strategies for urban and rural areas is crucial for promoting low-carbon construction worldwide. Facilitating construction 5.0 for resilient, sustainable and smart cities is shown in Fig. 9.3.

9.5 Potential Regional Applications of UDTs and Construction 5.0 in Çatalköy–Esentepe

The Çatalköy–Esentepe area faces challenges in managing rapid tourism-driven urban growth, fragmented infrastructure systems, and pressure on coastal and energy resources. These issues provide a highly relevant context for the deployment of UDTs and Construction 5.0 technologies. Regionally grounded applications within the specific socio-economic and environmental conditions of Northern Cyprus are suggested as follows:

- **Energy Monitoring**: UDTs could support real-time simulation and monitoring of electricity usage across hotels, residential areas, and public buildings in Çatalköy–Esentepe. By integrating data from smart meters and solar installations, a local digital twin could help optimize grid loads, plan energy-efficient building retrofits, and align with North Cyprus's renewable energy goals.
- **Infrastructure Planning**: The region's ongoing infrastructure expansion—including road networks, utility systems, and housing—could be guided by UDT-based scenario modeling. A digital twin of the built environment could simulate the long-term impacts of zoning decisions, inform risk mitigation for areas exposed to coastal erosion or landslide hazards, and improve coordination between public authorities and private developers.
- **Tourism Flow Management**: As a fast-developing coastal destination, Çatalköy–Esentepe experiences seasonal surges in population. UDTs could assist in managing tourism pressure by modeling pedestrian and traffic flows, monitoring environmental stress in sensitive areas (e.g., beaches, historical sites), and supporting dynamic urban services such as waste collection and water usage.
- **Construction 5.0 Applications**: Within the framework of Construction 5.0, advanced digital tools and human-centric automation could be piloted in new construction projects in the region. For instance, modular building methods, robotics, and real-time data integration could improve the speed, safety, and environmental footprint of new residential developments, aligning with regional sustainability goals.

9.6 Conclusion

Urban resilience is the capacity of urban systems, including socio-ecological and socio-technical networks, to maintain or quickly restore functions and adapt to changes. It encompasses four pillars: recovery, adaptation, transformation, and resilience. Urban resilience can be viewed from engineering, ecological, and socio-ecological perspectives. Smart cities and resilient cities share development trajectories, though not all smart cities are resilient to unexpected events. Urban areas face challenges from rapid urbanization, inadequate planning, and natural hazards, particularly in emerging economies. Resilience and sustainability are critical strategies for improving urban planning and addressing hazards, highlighted in the Sustainable Development Goals (SDG 11). Understanding resilience involves exploring its theoretical evolution and interdisciplinary applications.

Integrating UDT into smart city frameworks has the potential to revolutionize urban development. The applications of UDT in urban planning are diverse, offering innovative solutions for energy management, infrastructure monitoring, and public services, among other areas. Many cities are exploring the concept of UDTs to enhance urban administration, development, and decision-making processes. While a fully realized city-scale UDT has not yet been achieved, there is a clear movement toward this goal. Even though urban digital twins are still in the early stages of development, they can already contribute to improvements in urban management. A city digital twin allows for the integration of city planning and management into a single platform, eliminating the need for multiple management systems. UDT can enable faster response times through autonomous actions and enhance city management efficiency by providing real-time insights into the city's condition.

Sustaining resilience in the AEC industry aligns with environmental responsibility and Construction 5.0's vision of smart, sustainable structures. This includes adopting sustainable techniques, decentralizing production, and large-scale customization. Success relies on balancing economic, environmental, and social aspects, with blockchain aiding sustainability and traceability. Society 5.0 emphasizes energy efficiency, automation, and occupant comfort in smart built environment.

References

Ahvenniemi H, Huovila A, Pinto-Seppa I, Airaksinen M (2017) What are the differences between sustainable and smart cities? Cities 60:234–245

Almusaed A, Yitmen I (2023) Architectural reply for smart building design concepts based on artificial intelligence simulation models and digital twins. Sustainability 15(6):4955

Alva P, Biljecki F, Stouffs R (2022) Use cases for district-scale urban digital twins. International Archives of the Photogrammetry. Remote Sensing and Spatial Information Sciences—ISPRS Archives

Apostu SA, Vasile V, Vasile R, Rosak-Szyrocka J (2022) Do smart cities represent the key to urban resilience? Rethinking urban resilience. Int J Environ Res Public Health 19:15410

Banihashemi S, Sohi SZ (2022) Data-centric regenerative built environment: big data for sustainable regeneration. Routledge

Batty M (2008) The size, scale, and shape of cities. Science (New York, N.Y.) 319(5864):769–771

Bibri SE (2021a) Data-driven smart sustainable cities of the future: an evidence synthesis approach to a comprehensive state-of-the-art literature review. Sustain Futures 3:100047

Bibri SE (2021b) Data-driven smart eco-cities and sustainable integrated districts: a best-evidence synthesis approach to an extensive literature review. Eur J Futures Res 9(1):16

Bibri SE (2023) Data-driven smart eco-cities of the future: an empirically informed integrated model for strategic sustainable urban development. World Futures 79(7–8):703–746

Bibri SE, Krogstie J (2021) A novel model for data-driven smart sustainable cities of the future: a strategic roadmap to transformational change in the era of big data. Future Cities Environ 7(1):1–25

Boyce DE, Williams HCWL (2015) Forecasting urban travel: past, present and future. Edward Elgar Publishing

Brand FS, Jax K (2007) Resilience as a descriptive concept and a boundary object. Ecol Soc 12:art23

Caprari G, Castelli G, Montuori M, Camardelli M, Malvezzi R (2022) Digital twin for urban planning in the green deal era: a state of the art and future perspectives. Sustainability (Switzerland) 14(10):6263

Chourabi H, Nam T, Walker S, Gil-Garcia JR, Mellouli S, Nahon K et al (2012) Understanding smart cities: an integrative framework. In: 2012 45th Hawaii international conference on system sciences. IEEE, pp 2289–2297

da Silva CA, dos Santos EA, Maier SM, da Rosa FS (2019) Urban resilience and sustainable development policies. Rev Gestão 27:61–78

De Falco S (2018) Geographic determinism Vs urban resilience: Italian scenario analysis. TeMA-J Land Use Mobil Environ 2018(11):65–88

Deng T, Zhang K, Shen ZJM (2021) A systematic review of a digital twin city: A new pattern of urban governance toward smart cities. Journal of Management Science and Engineering 6(2):125–134

Depretre A, Jacquinod F, Mielniczek A (2022) Exploring digital twin adaptation to the urban environment: comparison with cim to avoid silo-based approaches. ISPRS Annals of the Photogrammetry, Remote Sensing and Spatial Information Sciences

Deren L, Wenbo Y, Zhenfeng S (2021) Smart city based on digital twins. Comput Urban Sci 1:1–11

Elsehrawy R, Kumar B, Watson R (2021) A digital twin uses classification system for urban planning & city infrastructure management. J Inf Technol Constr 26:832–862

Ferré-Bigorra J, Casals M, Gangolells M (2022) The adoption of urban digital twins. Cities 131:103905

Fitzgibbons J, Mitchell C (2019) Just urban futures? Exploring equity in "100 Resilient Cities". World Dev 2019(122):648–659

Gholami H, Abu F, Lee JKY, Karganroudi SS, Sharif S (2021) Sustainable manufacturing 4.0—pathways and practices. Sustainability 13:13956

Hao H, Bi K, Chen W, Pham TM, Li J (2023) Towards the next generation design of sustainable, durable, multi-hazard resistant, resilient, and smart civil engineering structures. Eng Struct 277:115477

Holling CS (1973) Resilience and stability of ecological systems. Annu Rev Ecol Syst 1(4):1–23

Ikudayisi AE, Chan AP, Darko A, Yomi MD (2023) Integrated practices in the Architecture, Engineering, and Construction industry: Current Scope and Pathway towards Industry 5.0. J. Build Eng 73:106788

Ivanov S, Nikolskaya K, Radchenko G, Sokolinsky L, Zymbler M (2020) Digital twin of city: concept overview. In: Proceedings—2020 global smart industry conference. GloSIC 2020

Lei B, Janssen P, Stoter J, Biljecki F (2023) Challenges of urban digital twins: A systematic review and a Delphi expert survey. Automation in Construction 147, Article 104716. https://doi.org/10.1016/j.autcon.2022.104716

Lloyds (2017) Future cities: building infrastructure resilience. Report. Lloyds/Arup, London

LopezDeAsiain M, Díaz-García V (2020) The importance of the participatory dimension in urban resilience improvement processes. Sustainability 12:7305

Lu Q, Parlikad AK, Woodall P, Don Ranasinghe G, Xie X, Liang Z, Konstantinou E, Heaton J, Schooling J (2020) Developing a digital twin at building and city levels: case study of West Cambridge Campus. J Manag Eng 36(3):05020004

Major P, Li G, Hildre HP, Zhang H (2021) The use of a data-driven digital twin of a smart city: a case study of Ålesund, Norway. IEEE Instrum Meas Mag 24(7):39–49

Mashaly M (2021) Connecting the twins: a review on digital twin technology & its networking requirements. Procedia Comput Sci 184:299–305

McGill R (2020) Urban resilience—an urban management perspective. J Urban Manag 9:372–381

Meerow S, Stults M (2016) Comparing conceptualizations of urban climate resilience in theory and practice. Sustainability 8:701

Meerow S, Newell JP, Stults M (2016) Defining urban resilience: a review. Landsc Urban Plan 147:38–49

Miller TR (2013) Constructing sustainability science: emerging perspectives and research trajectories. Sustain Sci 8(2):279–293

Mohammadi N, Taylor J (2019) Devising a game theoretical approach to enable smart city digital twin analytics. In: Proceedings of the 52nd Hawaii International Conference on System Sciences, 2019

Monstadt J, Schmidt M (2019) Urban resilience in the making? The governance of critical infrastructures in German cities. Urban Stud 56:2353–2371

Murtagh N, Scott L, Fan J (2020) VSI editorial—sustainable and resilient construction: current status and future challenges. J Clean Prod 268:122264

Mylonas G, Kalogeras A, Kalogeras G, Anagnostopoulos C, Alexakos C, Muñoz L (2021) Digital twins from smart manufacturing to smart cities: a survey. IEEE Access Pract Innov Open Solut 9:143222–143249

Nochta T, Wan L, Schooling JM, Parlikad AK (2021) A socio-technical perspective on urban analytics: the case of city-scale digital twins. J Urban Technol 28(1–2):263–287

Papyshev G, Yarime M (2021) Exploring city digital twins as policy tools: a task based approach to generating synthetic data on urban mobility. Data Policy 3(5):e16

Peldon D, Banihashemi S, LeNguyen K, Derrible S (2024) Navigating urban complexity: the transformative role of digital twins in smart city development. Sustain Cities Soc 111:105583

Petrova-Antonova D, Ilieva S (2019) Methodological framework for digital transition and performance assessment of smart cities. In: 2019 4th international conference on smart and sustainable technologies. SpliTech 2019

Pirlone F, Spadaro I, Candia S (2020) More resilient cities to face higher risks. The case of Genoa. Sustainability 12:4825

Redman CL (2014) Should sustainability and resilience be combined or remain distinct pursuits? Ecol Soc 19:37

Roostaie S, Nawari N (2022) The DEMATEL approach for integrating resilience indicators into building sustainability assessment frameworks. Build Environ 207:108113

Salas J, Yepes V (2019) MS-ReRO and D-ROSE methods: assessing relational uncertainty and evaluating scenarios' risks and opportunities on multi-scale infrastructure systems. J Clean Prod 216:607–623

Sarker MNI, Wu M, Alam GMMGM, Shouse RC (2020a) Administrative resilience in the face of natural disasters: empirical evidence from Bangladesh. Pol J Environ Stud 29:1825–1837

Sarker MNI, Peng Y, Yiran C, Shouse RC (2020b) Disaster resilience through big data: way to environmental sustainability. Int J Disaster Risk Reduct 51:101769

Sarker MNI, Wen J, Yang B, Yusufzada S, Huda N, Mahbub F (2021) Assessment of environmental governance in disaster vulnerability context of rural bangladesh. Growth Chang 52:1155–1171

Shahat E, Hyun CT, Yeom C (2021) City digital twin potentials: a review and research agenda. Sustainability 13(6):3386

Shirowzhan S, Tan W, Sepasgozar SME (2020) Digital twin and CyberGIS for improving connectivity and measuring the impact of infrastructure construction planning in smart cities [Review]. ISPRS Int J Geo Inf 9(4):240

Tzioutziou A, Xenidis Y (2021) A study on the integration of resilience and smart city concepts in urban systems. Infrastructures 6:24

Wan L, Nochta T, Schooling JM (2019) Developing a city-level digital twin-Propositions and a case study. In: International conference on smart infrastructure and construction 2019, ICSIC 2019: driving data-informed decision-making, pp 187–193

Wang SM, Vu LH (2023) The integration of digital twin and serious game framework for new normal virtual urban exploration and social interaction. Journal of Urban Management 12(2):168–181

Weil C, Birbi S, Longchamp R, Golay F, Alahi A (2023) Urban Digital Twins: Challenges & Perspectives for Sustainable Smart Cities. Available at SSRN 4429379

Yang S, Kim H (2021) Urban digital twin applications as a virtual platform of smart city. Int J Sustain Build Technol Urban Dev 12(4):363–379

Yitmen I, Alizadehsalehi S (2021) Towards a digital twin-based smart built environment. In: BIM-enabled cognitive computing for smart built environment, pp 21–44

Yitmen I, Almusaed A, Alizadehsalehi S (2023) Investigating the causal relationships among enablers of the construction 5.0 paradigm: integration of operator 5.0 and society 5.0 with human-centricity, sustainability, and resilience. Sustainability 15(11):9105

Yitmen I, Almusaed A, Alizadehsalehi S (2024). Facilitating construction 5.0 for smart, sustainable and resilient buildings: opportunities and challenges for implementation, smart and sustainable built environment

Zhang X, Li H (2018) Urban resilience and urban sustainability: what we know and what do not know? Cities 72:141–148

Urban Resilience in Smart Cities: Lessons and Insights from a Global Perspective

10

Abstract

This chapter presents a comprehensive analysis of sustainable urban resilience development initiatives, with a particular focus on their implementation across European cities. It explores municipal commitments to advancing environmental sustainability and public health, alongside efforts to strengthen local governance frameworks aligned with the Sustainable Development Goals (SDGs). Key areas of focus include enhancing financial mechanisms for sustainable urban growth, fostering multi-stakeholder partnerships and co-creation processes, and instituting Low Emission Zones (LEZs) as strategic interventions for improving air quality. The chapter also examines the integration of green infrastructure and nature-based solutions as part of broader urban ecological strategies. It synthesizes lessons learned from past urban resilience initiatives, identifying critical insights for shaping future sustainable urban development. A case study of Moscow illustrates the role of green construction in promoting resilience, supported by urban planning regulations and risk-informed land-use strategies. Additionally, the chapter reviews emerging green construction standards and environmental certification systems in Russia, contextualized within global best practices. Beyond the European context, the chapter incorporates evidence from China, highlighting the role of smart city frameworks in advancing urban resilience. It analyzes how digital technologies contribute to scenario planning, infrastructure optimization, and adaptive governance, emphasizing the structural, technological, and distributive effects of smart city development. India's Smart Cities Mission (SCM) is also discussed, with emphasis on the integration of disaster resilience principles, participatory governance, and infrastructural preparedness. The chapter further synthesizes comparative findings from diverse urban settings, examining the interconnections between smart city strategies, disaster mitigation efforts, and urban resilience outcomes. It identifies key success factors underpinning the SCM, including institutional robustness, citizen engagement, and sustainable infrastructure design. A comparative evaluation of smart city business models in London, Amsterdam, and Berlin provides additional insights into financing mechanisms, public-private partnerships, and community-centered urban innovation.

10.1 Introduction

Considerable initiatives are underway in European cities to improve their urban landscape, aiming to provide sustainable and healthy living conditions for all residents. Nonetheless, cities

V. Yorucu et al., *Modern Approach of Resilient and Sustainable Smart Cities*, Sustainable Development Goals Series, https://doi.org/10.1007/978-3-032-05715-0_10

face a range of crises stemming from both natural and human-induced hazards. Natural hazards encompass floods, earthquakes, heatwaves, volcanic eruptions, and epidemics, while human-induced hazards include power outages, toxic material spills, and nuclear disasters associated with industrial incidents. These crises significantly affect local communities and living standards (Egegård et al. 2024; García-Blanco et al. 2023; Salama et al. 2024).

Cities have faced numerous crises in the past and have demonstrated effective recovery strategies. However, recent trends indicate an increase in the frequency and severity of events related to climate change, as well as other crises such as conflicts and wars. Recent research has focused on three main aspects: (a) developing frameworks and indicators to measure resilience (Clarke et al. 2020; Abusaada and Elshater 2023a; Osei-Kyei et al. 2023), (b) analyzing specific case studies of cities (Peres et al. 2017; Slaev and Nedovic-Budic 2017), and (c) examining particular types of crises within specific contexts (Majewska et al. 2022; Abusaada and Elshater 2023b), or concentrating on sustainability at a micro-level (Dessouky et al. 2023a, b).

Urban resilience, as described by the Resilient Cities Network (RCN 2022), refers to the inherent capacity of various elements within a city—such as systems, businesses, institutions, communities, and individuals—to withstand, adapt to, and thrive despite ongoing pressures and sudden disturbances. Similarly, the UN-Habitat defines urban resilience as "the measurable ability of any urban system, along with its inhabitants, to maintain continuity in the face of all shocks and stresses, while also positively adapting and evolving towards sustainability" (UN-Habitat 2016). This notion emphasizes the need for cities to adopt a comprehensive understanding of their strengths and vulnerabilities, encouraging inclusive participation from all segments of society, including the most marginalized. Ultimately, urban resilience embodies a proactive and holistic approach to urban development, aimed at enhancing the city's resilience to adversity, facilitating transformation, and fostering sustainable progress (Gkontzis et al. 2024; Salama et al. 2024).

Cities must proactively maintain a state of readiness, demonstrated through three phases: preparation, prevention, and action. This approach helps predict the extent of damage and enables swift recovery with minimal economic, environmental, and social consequences. European cities are particularly vulnerable to future shocks related to climate change, energy scarcity, and population growth. Addressing these challenges requires effective decision-making, governance, communication, and coordination mechanisms integrated into city organizational structures. However, merely reacting to crises is inadequate; cities must proactively plan for urban resilience (European Commission 2014; Matthies and Menne 2009). Sustainable strategies across all developmental activities aim to enhance overall quality of life. Developing greater urban resilience requires integrating horizontal and vertical frameworks and embedding urban resilience strategies at all spatial scales (Manca et al. 2017). To enhance urban resilience, cities must adopt urban planning and building design strategies that enable them to respond and adapt to the physical, economic, and social challenges posed by climate change. Urban resilience is a critical component of the UN Sustainable Development Goals (SDGs), with cities playing a central role in their achievement. However, planning and implementation approaches for resilience at the city level remain fragmented and limited (RCN 2022).

Urban environments are complex systems that require innovative solutions to address their challenges. The adoption of digital technologies has emerged as a significant catalyst in managing modern cities. The concept of smart cities has evolved, aiming to create sustainable and citizen-centered urban environments. This transformation relies on advanced technologies to improve various aspects of urban life, including infrastructure and public services. Central to smart city initiatives is the integration of data analysis, predictive modeling, and digital twin technologies, particularly at the neighborhood level

(Adreani et al. 2024). The key components of these technologies were explored to improve urban efficiency, bolster resilience, and elevate residents' quality of life (Deng et al. 2021; Weil et al. 2023).

This chapter discusses how sustainable urban resilience development initiatives were implemented by European cities, cities' commitment to Europe's environmental and public health, local governance for sustainable development, enhancing financial support for sustainable urban growth, partnerships and co-creation for sustainable development, low emission zones (LEZS) for clean air as part of sustainable development measures, green solutions and bringing nature to the city, lessons learned for future urban resilience, lessons learned for a future sustainable urban development, the case of Moscow involving green construction and urban sustainable development planning for city resilience, lessons from international experience involving the emerging green construction standards and certification systems in Russia, evidence from China involving achieving resilience through smart cities, construction of smart city involving new mode of urban development and three effects of enhancing urban resilience, India's smart cities mission including disaster resilience, adapting lessons from urban resilience initiatives, comparative analysis of smart cities, mitigation, and disaster resilience, key success factors of the Smart Cities Mission (SCM), interconnections between SCM, disaster mitigation, and disaster resilience, business models used in smart cities, and comparison of the Case Studies in London, Amsterdam and Berlin.

10.2 Sustainable Urban Resilience Development Initiatives Implemented by European Cities

Assessing cities' efforts toward sustainable development involves linking key themes related to urban resilience and sustainability with recent reports and policy documents from various European networks such as Diesel Net Europe, EUROCITIES Network, European Green Capitals Network, Urban Access Regulation in Europe, and WHO European Healthy Cities Network, as well as global networks like ICLEI–International Council for Local Environmental Initiatives and UIA–International Union of Architects, Commission on the UN Sustainable Development Goals (Eurocities 2025; European Environment Agency 2024; ICLEI 2025; Resilient Cities Network 2025; Salama et al. 2024).

10.2.1 A Commitment to Europe's Environmental and Public Health

Many European municipalities are committed to promoting healthy, sustainable development, and environmental-social resilience, as demonstrated by their active participation in global urban sustainability networks. These networks, as evidenced in recent literature, address various aspects of resilience and emergency readiness. Notable examples include the European Healthy Cities Network (EHCN), which prioritizes health as a key indicator of urban resilience through initiatives like "Healthier and Happier Cities for All," based on the 6P concept: "people, participation, prosperity, planet, place, and peace." The European Healthy Cities Network Charter, stemming from the International Healthy Cities Conference, introduces new health responsibilities for mayors, aligns Sustainable Development Goals (SDGs) with network objectives, and highlights the impact of healthy cities (WHO EHCN 2018, 2019). Moreover, the EUROCITIES Network, comprising approximately 190 cities, has developed strategic frameworks covering various quality-of-life dimensions. These frameworks prioritize people as primary stakeholders and advocate for inclusive societies, prosperous local economies, healthy living environments with vibrant public spaces, and assist city governments in addressing global challenges (Eurocities 2014). Notably, ICLEI—Local Governments for Sustainability, established in 1990, emphasizes SDG localization through "15

Pathways to Localize the Sustainable Development Goals" (ICLEI 2025). It provides resilience-enhancing tools like the European Resilience Management Guideline and advises on fostering resilient cities by integrating nature and addressing challenges such as resource overuse, fragile infrastructure, and social cohesion. The study evaluating the impact of the European Healthy Cities Network on Urban Resilience examines how the European Healthy Cities Network (EHCN) has influenced urban resilience by prioritizing health as a key indicator. It analyzes initiatives like "Healthier and Happier Cities for All," based on the 6P concept—people, participation, prosperity, planet, place, and peace—and assesses their effectiveness in enhancing urban resilience. Commitment to Europe's environmental and public health is illustrated in Fig. 10.1.

10.2.2 Local Governance for Sustainable Development

Enhancing sustainability-focused local governance entails bolstering leadership in Sustainable Development Goals (SDG), multilevel governance, and capacity building across national, regional, and local levels through collaboration with key stakeholders. Referring to the EUROCITIES study on Sustainable Development Objectives at the Local Level (EUROCITIES 2020), two primary actions facilitate this integration. Firstly, it involves incorporating SDGs into city strategies, enabling comprehensive sustainable development at the city level, as evidenced by strategic alignment and dedicated coordination mechanisms observed in various cities such as Bristol, Copenhagen, Ghent, and Gothenburg. Secondly, it requires enhancing leadership within public administration and local capacity, which is crucial for sustainable urban development. This is exemplified by strategies such as fostering new visions and leadership on SDGs within local administration (Strasbourg), conducting sustainability training for personnel (Gothenburg, Malmö), and implementing capacity building programs for municipal responsibilities (Bonn) (Salama et al. 2024; Kramer et al. 2024). Local governance for sustainable development is exhibited in Fig. 10.2.

10.2.3 Enhancing Financial Support for Sustainable Urban Growth

Increasing awareness to encourage businesses to engage in advancing Sustainable Development Goals (SDGs) and establishing platforms for coordination on sustainable development are measures aimed at mitigating risks associated with investments in sustainable innovations, as

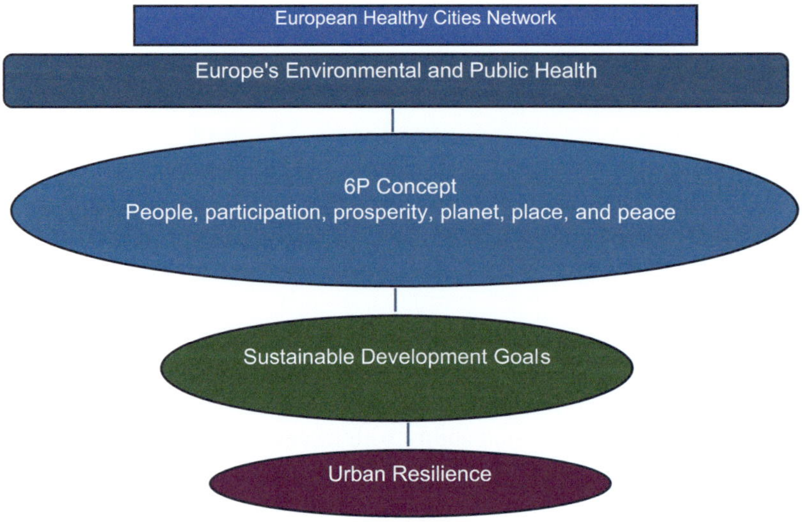

Fig. 10.1 Commitment to Europe's environmental and public health. (Source: Author's compilation)

Fig. 10.2 Local governance for sustainable development. (Source: Author's compilation)

part of initiatives to enhance financing for sustainable urban development (EUROCITIES 2020). In this context, two key practices in cities can be identified. Firstly, integrating SDGs into city budget processes, encompassing resource allocation, investment mobilization, participatory budgeting, and addressing interconnected issues. Cities leading in this practice include Gothenburg, Malmö, Strasbourg, Mannheim, and others that have adopted innovative financial forecasting models and incorporated climate criteria into municipal council submissions, such as Cologne and Stuttgart. Secondly, prioritizing sustainability in procurement involves integrating social and environmental considerations into procurement procedures. Various approaches have been adopted by European cities, including implementing new sustainable and equitable procurement legislation (Copenhagen, Oslo), utilizing multi-dimensional evaluation frameworks to assess service and product lifespans (Oslo), and establishing eco-fair trade and socially responsible public procurement measures (Hamburg, Stuttgart) (Salama et al. 2024). Figure 10.3 shows enhancing financial support for sustainable urban growth.

10.2.4 Partnerships and Co-creation for Sustainable Development

Goal 17 of the Sustainable Development Goals (SDGs), which centers on partnerships, serves as a crucial framework for realizing the 2030 Agenda and all other SDGs. This goal recognizes the importance of innovative, experimental, and unconventional partnerships alongside traditional multi-stakeholder collaborations. According to Eurocities (2020), numerous cities are implementing procedures and policies to address the Sustainable Development Goals as an integral component of sustainable development. These initiatives include supporting local partnerships and co-creation, fostering ideas for creating sustainable, livable cities with civic participation (Vienna), establishing multi-stakeholder alliances (Bristol, Mannheim) involving civil society, and forming consultative bodies comprised of civil society members (Mannheim) (Salama et al. 2024).

10.2.5 Low Emission Zones (LEZs) for Clean Air as Part of Sustainable Development Measures

Considering the air pollution crisis, European cities have introduced Low Emissions Zones (LEZs) as a measure to meet the EU Air Quality Standards and minimize exposure to air pollution (EUROCITIES 2020). Although LEZ regulations vary across the EU, common measures include: (a) restricting higher emission vehicles, (b) retrofitting and improving vehicle emissions, and (c)

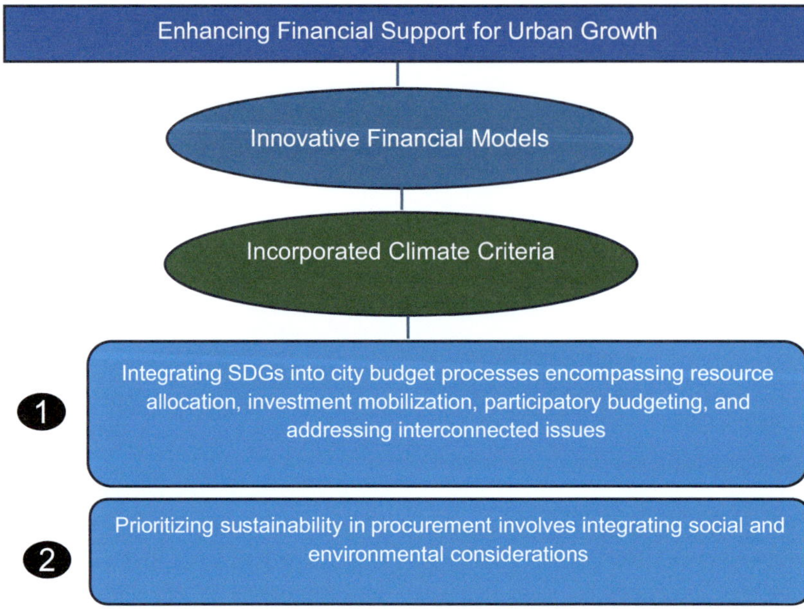

Fig. 10.3 Enhancing financial support for sustainable urban growth. (Source: Author's compilation)

implementing charging schemes. Additionally, cities are working towards transforming LEZs into zero-emission mobility zones (ZEZ) or Ultra Low Emission Zones (ULEZ) to complement policies promoting clean alternatives such as walking and cycling (e.g., London) and the electrification of all transportation modes, including public transit, taxis, shared and private vehicles, and delivery vehicles (Salama et al. 2024).

10.2.6 Green Solutions and Bringing Nature to the City

The European Green Capitals Network (European Commission 2014) recognizes cities' efforts to foster sustainable environments and communities through an award program. This initiative aims to share best practices and develop solutions for urban sustainability in Europe, while also representing cities in relevant global and European political processes. By embracing nature-inspired solutions, cities enhance their resilience and reap environmental, social, and economic benefits. EGCN highlights challenges related to the integration of built and natural environments, such as resource depletion affecting citizens' well-being, vulnerable infrastructure, and the importance of social cohesion for resilience. European cities tackle these issues through various initiatives, including waterway restoration (e.g., Essen), nature-based and technical stormwater management (e.g., Lahiti), adoption of the Green Space Index to assess ecosystem services (e.g., Stockholm), and the promotion of green roofs and corridors (e.g., Vitoria-Gasteiz) (Salama et al. 2024). Table 10.1 summarizes the key sustainable urban resilience initiatives in European cities.

10.3 Lessons Learned for Future Urban Resilience and Future Sustainable Urban Development

10.3.1 Lessons Learned for Future Urban Resilience

Analyzing 11 crisis response experiences and reviewing urban resilience strategies of ten Resilient Cities Network members reveals crucial lessons. Key aspects include communication, coordination, governance, social dimensions,

Table 10.1 Key sustainable urban resilience initiatives in European cities

Initiative	Description	Examples of cities
Commitment to Environmental and Public Health	Participation in sustainability networks like EHCN and EUROCITIES to promote urban resilience and public health	London, Copenhagen, Gothenburg
Local Governance for Sustainability	Integration of SDGs into city strategies, leadership development, and administrative capacity-building	Bristol, Strasbourg, Malmo
Financial Support for Sustainable Urban Growth	SDG-based budgeting, participatory funding, and sustainable procurement policies	Cologne, Oslo, Hamburg
Partnerships and Co-Creation	Multi-stakeholder alliances and civic participation for sustainable city development	Vienna, Mannheim, Bristol
Low Emission Zones (LEZs)	Restrictions on high-emission vehicles, promoting zero-emission transport alternatives	London, Stuttgart, Oslo
Green Solutions and Nature-Based Approaches	Restoration of urban waterways, green roofs, and technical stormwater management	Stockholm, Essen, Vitoria-Gasteiz
Integrated Crisis and Resilience Strategies	Lessons from 11 crisis response experiences and urban resilience strategies of 10 Resilient Cities Network members	L'Aquila, Paris, Lisbon, Eyjafjallajökull, Steigen, Kos, Stockholm, London, Venice

Source: Author's compilation

participatory planning, community engagement, diverse urban initiatives for sustainable development, and a commitment to health and resilience strategies (Salama et al. 2024).

1. *Communication and Coordination*
   Cities face significant challenges in crisis response due to coordination, communication issues, and siloed work at the administrative level. Paris floods and heatwaves, and L'Aquila earthquakes, highlight how multiple administrative units hindered effective communication and coordination (WHO 2022a, b; Salama et al. 2024). In L'Aquila, national-level decisions marginalized local authorities, hampering risk control and crisis response. Clear, effective communication methods are vital to coordinating national and local levels, ensuring efficient crisis management in urban resilience planning.
2. *Governance and Decision Making*
   Cases explored highlighted government efforts involving new local laws, regulations, and committees to prevent crises. The Paris floods, heatwaves, L'Aquila's earthquake, and Ajka's sludge spill disaster showed national policies forming without local authority input, affecting practical action. Effective urban resilience requires integrated decision-making frameworks, vertically and horizontally, with clear roles and accountability. Localizing national actions by including local authorities in decisions and implementation is essential (European Commission 2014; Plummer et al. 2018; Salama et al. 2024).
3. *Social Dimension of Urban Resilience*
   In some cases, the social dimension has been neglected in crisis responses. For example, the response to the earthquake in L'Aquila was a linear technical problem-solving exercise, ignoring social and behavioral concerns from relocation. Psychological issues, broken family relationships, and lack of social cohesion were overlooked.
4. *Participatory Planning and Community Engagement*
   Measures proposed by communities or through participatory decision-making pro-

cesses strengthen social resilience. The River Seine floods in Paris saw flood storage site construction and hazard mitigation through dialogue and participatory processes. The EVA project in L'Aquila, a community-based ecovillage, fostered a sense of community and ownership. Lisbon's Municipality engaged local partnerships to monitor hazards, assess risks, and promote good practices, strengthening the city's resilience to natural disasters.

5. *Diverse Initiatives for Urban Resilience and Sustainable Development*
   Cities are implementing various initiatives for sustainable development. Paris focused on flood mitigation and sustainability strategies. Kos implemented emergency planning, port restoration, and health tourism promotion. Lisbon prioritized housing, mobility, sustainability, and public spaces. Steigen emphasized nature and quality of life. Stockholm and London addressed pandemic measures based on local conditions. Venice adopted green infrastructure, risk mapping, and integrated urban modeling to improve urban conditions.
6. *Commitment to Health and Resilience Strategies*
   Analyzing ten cities' resilience strategies reveals a commitment to health in design and planning processes. While resilience is explicitly stated in visions, strategies, and policies, many plans lack the rigor needed for needs assessment and performance-based measures. Although over 20 cities have established resilience strategies, it remains to be seen how these strategies are implemented beyond prescriptive measures.

These lessons underscore the importance of integrated, community-focused, and adaptable approaches in urban resilience planning. Lessons learned for future urban resilience is illustrated in Fig. 10.4.

10.3.2 Lessons Learned for a Future Sustainable Urban Development

The examination of academic literature and policy documents on European cities and global networks yielded critical insights. These insights encompass key themes, including urban sustainability initiatives, governance and financial strategies, participatory co-creation processes, public health considerations, emission reduction efforts, urban greening, and the implementation of sustainable development projects (Salama et al. 2024).

Fig. 10.4 Lessons learned for future urban resilience. (Source: Author's compilation)

1. *Repositioning Urban Strategies through Sustainable Development Initiatives: A Shift from Crisis Response to Resilient and Integrated Approaches*
   Cities are increasingly recognizing that traditional crisis response mechanisms are insufficient, prompting a shift towards resilience strategies that are integrated with Sustainable Development Goals (SDGs). Addressing climate change and pursuing SDGs have emerged as central urban priorities. Notable initiatives, such as London's low-traffic neighborhoods, Stockholm's eco-smart projects, Kos's promotion of health tourism, and targeted investment plans, highlight a holistic approach to fostering urban resilience and sustainable development.
2. *Governance and Finance for Sustainability: Strengthening Local Governance and Integrating SDGs into Urban Development*
   An evaluation demonstrates strong commitment to sustainability in local governance, with a focus on SDG achievement through coordinated actions across various governmental levels. In selected cities, funding for sustainable urban development is gaining momentum, particularly with the integration of SDGs into budgeting processes and sustainable procurement practices. Key strategies for effective implementation include enhancing procurement policies to address social and environmental factors, reinforcing leadership for SDGs, fostering multi-level governance, and building capacity with stakeholders.
3. *Co-Creation for Sustainable Development: Leveraging Local Partnerships and Innovative Alliances*
   Several European cities are embracing co-creation as a mechanism for sustainable urban development, involving local partnerships and active engagement in decision-making. Achieving the full potential of the 2030 Agenda requires not only traditional multi-stakeholder partnerships but also innovative, non-traditional alliances. Local authorities stand to benefit from co-creating policy solutions and implementing participatory techniques within both policy and project domains.
4. *Health, Low/Zero Emission Zones, and Urban Greening: Promoting Clean Air and Well-being through Strategic Initiatives*
   Cities are committed to reducing pollution and ensuring clean air through the establishment of Low Emission Zones, strengthening health infrastructure via legislation, and implementing impactful public and green space initiatives that enhance well-being and quality of life. Efforts to achieve low or zero emissions are gaining traction, with a focus on pedestrianizing public spaces and promoting walkability. Urban greening initiatives, such as integrating water bodies into the cityscape and employing tools like the Green Space Index, aim to optimize the eco-efficient use of green spaces.
5. *Contextualized Sustainable Development Projects: Localizing SDGs through Collaborative and Interdisciplinary Approaches*
   Several projects focus on localizing and contextualizing SDGs within specific communities, highlighting efforts to utilize smart technologies for public benefit, establishing innovative governance systems, and engaging in participatory practices. These initiatives also promote circular economy models that reduce resource consumption and waste production. Such projects require interdisciplinary collaboration, involving engineers, public health experts, managers, administrators, architects, and designers, ensuring synergy across all sustainable development goals. Lessons learned for the future sustainable urban development is illustrated in Fig. 10.5.

10.4 Green Construction and Urban Sustainable Development Planning for City Resilience: The Case of Moscow

Green construction involves building new structures and renovating existing ones according to new standards that emphasize energy efficiency, protection against hazardous climate change

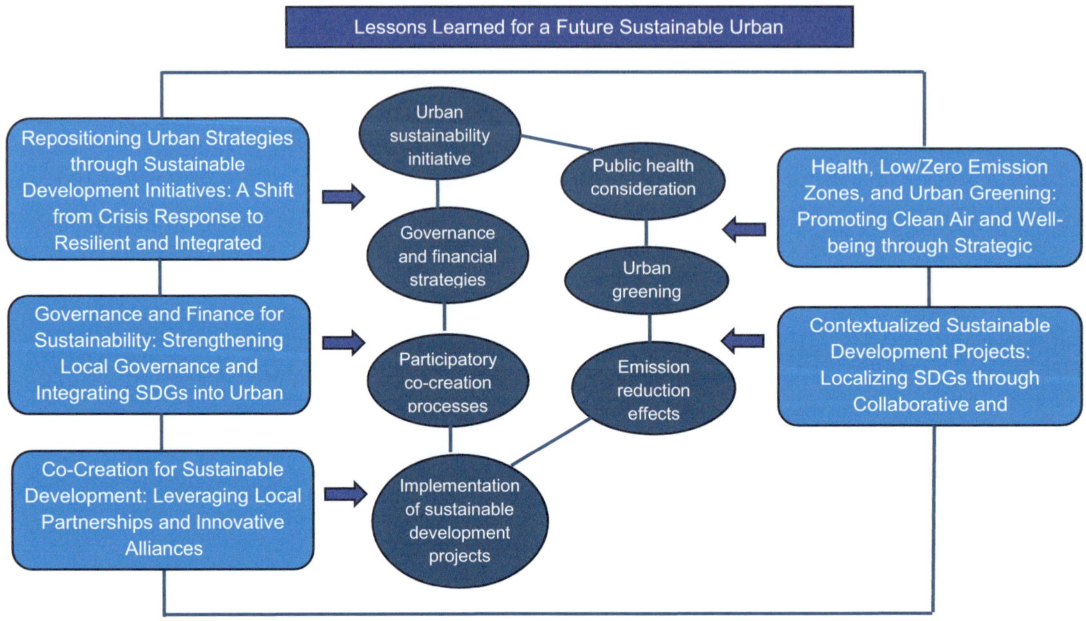

Fig. 10.5 Lessons learned for the future sustainable urban development. (Source: Author's compilation)

impacts, and creating comfortable living and working environments (Henderson et al. 2015, p. 2; UN-Habitat 2016, p. 17). This approach views buildings and structures as human environments that, aligned with the social pillars of sustainable development, must ensure comfortable conditions throughout their life cycle, from design to demolition. Economically and environmentally, buildings should be energy- and resource-efficient and minimize environmental impacts.

New urban planning, building regulations, and standards cover both individual buildings and the broader city territory. In the USA, federal building design regulations (International Building Code 2009) coexist with specific city rules, such as New York City's urban regulations. In Europe, EU directives and national regulations often take precedence. In Russia, the Urban Planning and Design Code of the Russian Federation (2004) sets out key organizational and legal provisions for urban development, used as practical management tools for sustainable urban development. In large cities like Moscow, local urban planning codes must align with federal regulations.

The Code of Practice mandates that city master plans include all necessary research, engineering calculations, and personnel training to ensure the resilience of buildings and infrastructure. This requires regional and local urban planning regulations tailored to specific urban areas, addressing aspects such as high-rise construction, area development, energy use, and building height. The primary tool for urban resilience and sustainable development is land use or functional zoning, which aligns with specific economic sectors and end users, including industrial, residential, recreational, environmental/disaster protection, or mixed-use areas. It also involves specific construction activities related to disaster preparedness and response, ensuring compliance with urban planning regulations.

In Moscow, resilience and environmentally sustainable development are central to "The Main Directions of Urban Development of Moscow," detailed in the "General Plan of Moscow Development by 2025" (2010). This strategic document outlines the city's long-term development priorities, ensuring sustainability and resilience by aligning environmental, social, and economic goals. Key components of the General

Plan include (1) defining areas at risk of emergencies from natural and man-made hazards and (2) delineating areas impacted by nearby capital construction facilities.

These priorities are implemented through specific projects focusing on green construction. Examples include transforming the former industrial ZIL car factory site into a business and residential area and replacing the demolished "Russia" hotel with an ecological park in central Moscow. The trend towards green construction is further evidenced by public rejection of large-scale projects, including those by renowned green architect Norman Foster, due to poor environmental integration.

Engineering and geological mapping of Moscow, focusing on construction risks like landslides, flooding, and pollution, has identified nine environmentally hazardous types of areas. This data has informed the risk assessment of 200 high-rise construction sites and led to the development of Temporary Urban Standards for High-Rise Buildings (MGSN 4-19-2005).

Green construction projects aim to reduce industrial NOx emissions by 80% by 2025, limit daily water consumption to 235 liters per person, and increase recycling of wastewater and municipal waste. These initiatives utilize low-waste, water-saving, and energy-saving technologies, reducing thermal energy use by 5–6% and fuel consumption by 20–25%. Specific projects include the construction, reconstruction, or elimination of 11 toxic waste-processing enterprises and 5 municipal waste incineration plants, handling up to 1.2 million tons annually. Figure 10.6 shows the green construction and urban sustainable development planning for city resilience.

Regional standards play a crucial role in urban planning and construction, addressing Moscow-specific issues more effectively than federal regu-

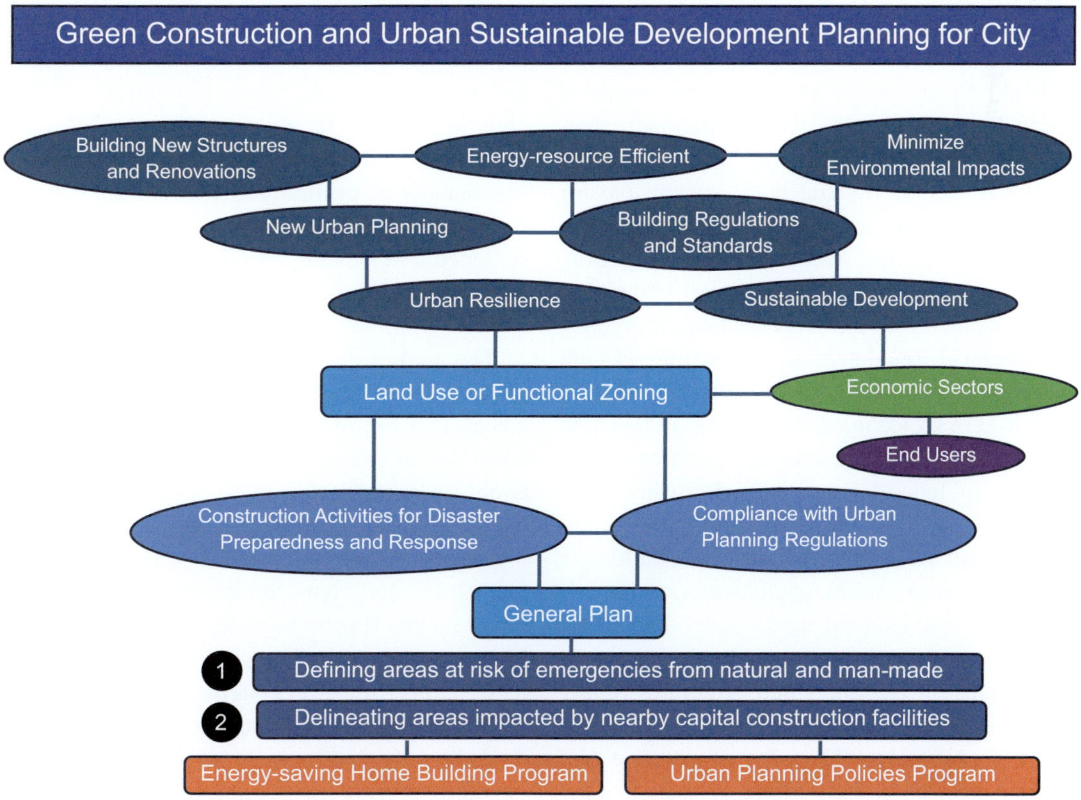

Fig. 10.6 Green construction and urban sustainable development planning for city resilience. (Source: Author's compilation)

lations. To streamline the process, Moscow authorities use city government resolutions to approve urban programs, such as the "Energy-Saving Home Building" program and the "Urban Planning Policies" program. These programs enhance building energy efficiency, translating the General Plan into concrete projects meeting economic, social, and environmental resilience criteria.

10.5 How International Experience is Shaping Russia's New Green Construction Standards and Certifications

The current design and construction standards for buildings and structures in Russia have been in effect since 2004, following the replacement of the earlier state standards system in accordance with the federal law on technical regulation. These standards align with the 2009 Technical Regulation on Buildings Safety and the 2009 Federal Law on Energy Saving and are regularly updated to match relevant international standards. The Technical Committee for Construction TK465 is the sole federal body responsible for reviewing all rules and standards, with approval shared between the Ministry of Construction, Housing and Utilities of the Russian Federation (which handles the rules) and the Federal Agency for Technical Regulation and Metrology (which oversees the standards). This is different from the USA, where six technical committees of business and research experts update and approve the six volumes of federal regulations (Porfiriev et al. 2017).

The qualitative characteristics of a building's design—ensuring safety of construction materials, waste management, and energy and water saving—are primarily governed by the Technical Regulations on Buildings and Structures Safety approved by federal law (Federal Law 2009a) and approximately 200 codes of practice and national standards. Fire safety is regulated by special technical regulations approved by Federal Law No. 123-FZ from July 22, 2008. According to these documents, building resilience involves eight dimensions: (1) mechanical safety (robustness); (2) fire safety; (3) safety in natural and technological emergencies or against hazardous impacts; (4) safe living conditions for human health; (5) safe building use by occupants; (6) accessibility for disabled people; (7) energy efficiency; and (8) minimal environmental impact.

The Urban Planning and Design Code requires the inclusion of environmental standards in building design, substantiated in the construction project documentation section for environmental impact assessment (Porfiriev et al. 2017). This uses criteria and indicators central to the federal 'green' standard, which regulates compliance assessment of real estate construction to respective environmental requirements (National Standard 2013) and the standards of the National Construction Workers Association. These standards remain voluntary for investors, designers, and inspection agencies, with their implementation being somewhat limited. However, Federal Law No. 385-FZ from December 30, 2009 (Federal Law 2009b), which amended existing technical regulations, allows for the use of foreign legislation in Russia. This enables the increasing application of international construction standards, primarily BREEAM and LEED, for assessing the environmental compliance of designed and existing buildings, leading to their certification as green edifices (Porfiriev et al. 2017).

The analysis of certification systems in construction worldwide indicates that energy efficiency is a top priority for both BREEAM and LEED, with scores of 30 and 33 points, respectively. In Russia, the green standard GOST R 54964-2012 takes a different approach to assessing energy saving and efficiency, utilizing seven indicators and ten minimum environmental requirements (Porfiriev et al. 2016). Figure 10.7 illustrates the emerging green construction standards and certification systems.

Regional urban planning regulations and standards for city development and implementation, including construction codes and codes of practice, are continuously updated and aligned with international standards. These standards increas-

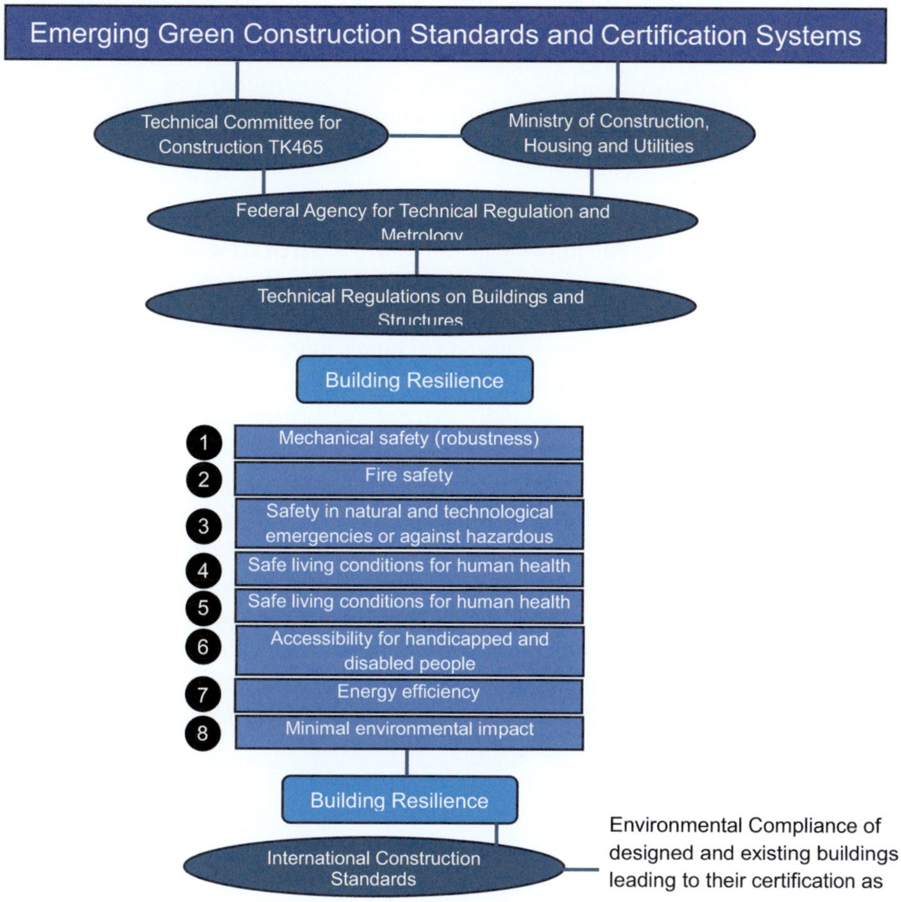

Fig. 10.7 Emerging green construction standards and certification systems. (Source: Author's compilation)

ingly emphasize resilience, robustness, energy efficiency, and environmental quality in buildings and structures. This alignment supports the green construction policy in Russia, which is carried out through federal, regional, and local targeted investment programs for the construction and reconstruction of buildings and structures.

The implementation of green certification and construction standards, both internationally and in Russia, is becoming a major policy tool for building resilient cities, which are essential for sustainable development at national, regional, and global levels. Within this green construction policy, energy saving and efficiency are prioritized, contributing significantly—over one-third of the total factor input—to enhancing the resilience and sustainability of cities (Porfiriev et al. 2017).

10.6 Achieving Resilience Through Smart Cities: Evidence from China

While China has been advancing in smart city development for several years, the concept of urban resilience is relatively new and is gaining increasing attention (Bai et al. 2019). The conflict between urban development and urban resilience has recently become a significant concern. For instance, the sudden outbreak of the Coronavirus disease (COVID-19) in Wuhan in January 2020 led to a lockdown of the city and severely disrupted the operations of nearly all cities in China by spring 2020 (Tong et al. 2020), highlighting a lack of urban resilience. In this context, effectively integrating urban resilience into the process of smart city development has become a matter of broad policy and practical importance (Zhou et al. 2021).

10.6.1 Smart Cities and Urban Resilience

The concept of urban resilience encompasses multiple dimensions. It can refer to the ability to minimize impact and harm during crises (Alberti 2005; Du et al. 2020) or the capability of an urban system to maintain its original functions and development during emergencies (Ribeiro and Pena Jardim Gonçalves 2019). Larger city sizes increase the vulnerability of urban systems to natural disasters and social crises (Kummitha and Crutzen 2017; Labaka et al. 2019). A key topic within this concept is inclusive and open urban resilience (Leitner et al. 2018). To support a multi-level urban system, urban leadership, and a diversified economic structure, innovation systems, and networks provide the necessary institutional background (Evans and Karecha 2014). The stronger a city's resilience, the higher its social well-being (Labaka et al. 2019). Smart cities significantly enhance urban resilience (Zhu et al. 2019), offering smart solutions for urban construction and addressing major social challenges such as overpopulation, transportation, pollution, sustainability, safety, health, and the creation of new businesses (Abella et al. 2017). Smart cities also have a higher propensity to apply for patents, particularly in high-tech fields, with technologies related to smart city policies having spillover effects (Caragliu and Del Bo 2019). However, the concept of smart cities is still somewhat controversial (Kitchin 2014), with most studies focusing on the role of government in smart city development (Hu and Zheng 2020), related policies (Caragliu and Del Bo 2019; Soyata et al. 2019), and technology development (Allam and Dhunny 2019; Soyata et al. 2019).

China's smart city model features a distinct layered governance structure, where the government's role is crucial in building smart infrastructure, environmental protection, and fostering smart technologies (Hu and Zheng 2020; Huang and Chan 2018). Consequently, despite the inevitability of smart cities in China's development, the disorderly growth of Chinese cities has been effectively controlled, and smart city development has progressed to a significant extent (Chen et al. 2018).

10.6.1.1 Smart City, Industrial Structure, and Urban Resilience

Smart cities utilize technologies like the Internet of Things, digitization, and remote sensing to enhance urban socio-economic development, thereby facilitating smart industry, governance, and improved livelihoods (Allam and Dhunny 2019; Ismagilova et al. 2019; Kitchin 2014). By integrating emerging technologies with traditional industries, they transform production methods, which consistently boost the efficiency of these industries, speed up their modernization, and achieve the optimization and upgrading of the industrial structure to a higher proportion of the tertiary sector (Cui and Yang 2008; Wang et al. 2020).

However, urban resilience varies greatly across regions (Martin et al. 2016), with industrial diversity playing a crucial role (Doran and Fingleton 2018; Evans and Karecha 2014). Cities with a high degree of industrial specialization and continuous concentration exhibit lower resilience due to a lack of flexibility in competitive and productive sectors. Conversely, cities with diversified industrial structures have more capacity and speed for industrial adjustments, making them more resilient (Fingleton and Palombi 2013; Mai et al. 2021).

From an economic cycle perspective, industrial diversity greatly enhances economic resilience during recessions. Regions with high industrial concentration perform better when unemployment is low, but during periods of high unemployment, industrial diversification proves more beneficial for urban resilience (Brown and Greenbaum 2017).

Smart city development generally enhances urban resilience but impacts various dimensions differently, including urban economic resilience, social resilience, ecological resilience, and infrastructure resilience. As a novel evolution of traditional city growth, smart cities create distribution, technological, and structural effects through innovation (Grossman and Krueger 1991; Shi et al. 2018). These effects should naturally lead to improvements in urban resilience.

10.6.2 Construction of Smart City: New Mode of Urban Development—Three Effects—Enhancing Urban Resilience

The three effects, as analyzed by Shi et al. (2018), are applied to the construction of smart cities and urban resilience development. The allocation effect is adjusted to the distribution effect for better alignment with this study, while the structural and technological effects remain unchanged.

The construction of a smart city enhances urban resilience through several key effects, including the structural effect.

1. *Structural Effect*: The structural effect from smart city development directly promotes industrial structure optimization, thereby continuously enhancing urban resilience. Emerging technologies in smart city construction accelerate the transition from old to new driving forces. This not only fosters high-tech and emerging industries but also aligns resource allocation with the current economic development "new normal" (Michaels et al. 2012). Government-led industrial upgrading can achieve significant transformation by replacing outdated elements with new ones, continuously driving economic transformation and upgrading (Jang and Li 2018). This economic development has become an internal force, helping cities address internal issues and external disasters.

   Smart city construction enhances urban resilience through technological advancements.
2. *Technological Effect*: The technological effect of smart city development significantly enhances urban resilience by promoting the aggregation, integration, and optimal allocation of innovative elements. Smart city development relies heavily on scientific and technological innovation, with improvements rooted in fostering a robust technological innovation system led by enterprises (Tommi 2018).

   On one hand, enterprises benefit from smart city construction by improving their innovation capabilities. This is achieved by creating an efficient, open innovation network through advanced information infrastructure that facilitates the sharing of knowledge and creativity (Lee et al. 2015). On the other hand, smart city construction gradually improves the urban institutional environment, allowing enterprises more time and resources to focus on innovation activities, thereby enhancing their innovative capacity (Baumol 1993).

   The widespread adoption of emerging technologies drives connectivity across social, economic, and infrastructure domains in smart cities (Margarita 2014). This technological integration ensures that smart cities are better equipped to handle various challenges, thus improving their overall resilience.

   Smart city construction enhances urban resilience through the distribution effect.
3. *Distribution Effect*: The distribution effect of smart city development aims to balance the interests of multiple stakeholders to achieve a win-win situation for various parties (Xia et al. 2020). By leveraging the Internet of Things, big data, and other next-generation information technologies, smart cities drive improvements in capital, labor, technology, and other resources, thereby boosting resource allocation efficiency and fundamentally transforming the economic model. This process ultimately meets the fundamental requirements of innovation and development, leading to better urban income distribution and improved wages for residents.

   On one hand, a more equitable distribution effect will attract a larger population concentration, providing momentum for the innovative development of smart cities (Wang and Yi 2018). On the other hand, improved income distribution and sustained economic growth will effectively reduce poverty (Bourguignon 2004), thereby enhancing urban resilience in economic, social, and other domains.

10.7 India's Smart Cities Mission to Include Disaster Resilience

This chapter examines the complex interplay between urban development, disaster mitigation, and resilience, highlighting the critical need to address disaster vulnerability in urban environments in India. Given the high population density, limited resilience, and inadequate coping capacities in many urban areas, a significant portion of the population remains at risk. The study underscores the importance of integrating disaster resilience principles into India's Smart Cities Mission, with a particular focus on infrastructure development, the implementation of early warning systems, and enhanced community engagement as key strategies for strengthening urban resilience (Meduri et al. 2024).

10.7.1 Adapting Lessons from Urban Resilience Initiatives

The Making Cities Resilient (MCR) Campaign, launched in 2010 by the United Nations Office for Disaster Risk Reduction (UNDRR 2024), introduced ten essential principles for fostering urban resilience. These principles serve as independent, yet interrelated steps aligned with the implementation of the Sendai Framework for Disaster Risk Reduction (2015–2030). The MCR campaign represents a cross-sectoral initiative aimed at achieving resilience through a three-pronged approach encompassing knowledge, planning, and implementation.

A 2019 study conducted by the Integrated Research and Action for Development (IRADe 2019), titled *Sustainable and Disaster Resilient Urban Development: Assessment of 10 Cities in India*, highlighted key areas requiring enhancement to strengthen disaster resilience. These areas include disaster risk reduction, database management, natural resource management via macro and regional planning, governance and institutional frameworks, technical capacity at the urban local body level, development of robust infrastructure, improvements in socio-economic conditions, and investments through public-private partnerships.

Additionally, Arup, in collaboration with the Rockefeller Foundation, developed the City Resilience Index (CRI), comprising 52 indicators categorized into 12 resilience goals across four dimensions: people, organization, place, and knowledge. This framework evaluates the robustness of infrastructure and ecosystems that protect urban areas from natural hazards. Achieving resilience in urban environments necessitates a multidimensional approach that addresses vulnerabilities across various sectors. Adopting and integrating strategies from the MCR campaign, IRADe assessments, and CRI methodologies can enhance the effectiveness of resilience-building measures.

10.7.2 Comparative Analysis of Smart Cities, Mitigation, and Disaster Resilience

The interrelationship between smart cities, disaster mitigation, and resilience can be conceptualized as interconnected mechanisms operating in a synchronized manner. Smart cities leverage advanced technologies to optimize planning and resource management, while mitigation strategies proactively reduce the impacts of potential hazards. Disaster resilience, in turn, functions as the foundational framework that enables cities to withstand and recover from adversities.

When these components interact effectively, urban areas become more capable of anticipating and mitigating risks through data-driven decision-making, resilient infrastructure, and community engagement. This integration fosters sustainability and security, ensuring that cities remain functional under both normal and extreme conditions. Furthermore, achieving disaster resilience necessitates equitable access to resources and support systems, recognizing diverse vulnerabilities, and ensuring targeted assistance during and after disaster events.

A resilient urban framework integrates disaster resilience within the Smart Cities Mission (SCM) by strengthening key infrastructural and social components. For instance, adaptable transportation networks facilitate swift evacuation and emergency response, a fortified healthcare system ensures medical aid availability during crises, and inclusive governance frameworks promote equitable resource distribution. Additionally, environmental sustainability measures, such as green infrastructure and disaster-conscious urban planning, play a crucial role in long-term resilience-building (Meduri et al. 2024). Table 10.2 shows the smart city resilience through core resilience domains.

10.7.3 Key Success Factors of the Smart Cities Mission (SCM)

The SCM in India aims to transform urban centers into more sustainable and livable environments through technology integration, infrastructure development, and citizen participation. The success of SCM can be evaluated across multiple dimensions, including:

1. *Infrastructure Development*—Improvement of transportation networks, water supply, sewage systems, and public spaces.
2. *Technology Integration*—Implementation of smart transportation systems, e-governance, and data analytics.
3. *Sustainability*—Reduction of carbon footprints and enhancement of environmental indicators.
4. *Quality of Life*—Provision of improved public services, congestion reduction, enhanced healthcare, and overall well-being.
5. *Citizen Engagement*—Active involvement of residents in governance and decision-making processes.
6. *Economic Growth*—Job creation, increased investments, and sustainable urban development.
7. *Replicability*—Scalability and adaptability of successful models across different urban settings.

10.7.4 Interconnections Between SCM, Disaster Mitigation, and Disaster Resilience

The integration of SCM, disaster resilience, and mitigation measures forms the foundation of a comprehensive urban safety strategy. These interconnected domains collectively enhance urban environments' ability to withstand and recover from hazards, ensuring sustainable development.

1. *Disaster Mitigation*—Encompasses structural measures (e.g., flood barriers, earthquake-resistant buildings) and non-structural measures (e.g., land-use planning, early warning systems) designed to minimize hazard impacts. The incorporation of mitigation principles in smart city planning enhances overall urban safety.
2. *Smart Cities Mission*—Focuses on architecture, urban planning, governance, and service management. By leveraging advanced planning methodologies, technology, and data-driven decision-making, SCM facilitates efficient disaster preparedness and response. Effective governance structures within SCM ensure that decision-makers have access to real-time, accurate information, which is crucial for disaster response and recovery.
3. *Disaster Resilience*—Comprises institutional, economic, infrastructural, social, and community resilience. SCM can contribute to disaster resilience by fostering inclusive governance, ensuring equitable access to resources, and promoting public awareness of disaster risks and preparedness measures. Additionally, the emphasis on environmental sustainability in smart cities supports ecosystem-based disaster resilience strategies, such as green infrastructure to mitigate flood risks and heat stress.

The synergy among these domains underscores the necessity of coordinated efforts to build safer,

Table 10.2 Smart city resilience through core resilience domains

Resilience domain	Related components
Structural resilience	– Architecture (ICT tools, Smart mobility, Water distribution, Health care) – Structural mitigation (hazard zoning, land use, site planning, natural ecosystems, material stipulation, construction codes, fire safety, plumbing services)
Social resilience	– Social inclusion – Educational equality – Health security – Transport access – Cultural diversity – Relationships between communities – Sense of community – Place attachment – Citizen participation
Economic resilience	– Equitable income – Female labor force participation – Diversified economic base – Employment (%) – House ownership (%) – Employment opportunities – Quality of life
Institutional resilience	– Planning and organizational systems – Strengthening local governance – Governance (logistics and supply chain, social inclusion) – Percentage population covered in mitigation
Infrastructural resilience	– Community response and recovery capacity – Percentage of shelters – Health care facilities – Road access – Waste management – Shelters for evacuation – Frequency of vulnerability assessment
Environmental resilience	– Water management – Pollution control – Housing – Air quality – Water resource management – Natural vegetation – Ecosystems and habitats
Knowledge and data resilience	– Integrated database management – Remote sensing data – Knowledge drivers (education, ICT knowledge, healthcare knowledge) – Prospective social innovation – % of knowledge workers
Services resilience	– Health services – Water services – Energy (generation, distribution, consumption)

Source: Author's compilation

more resilient cities. By integrating disaster mitigation principles within smart city initiatives, urban environments can achieve a balance between technological advancement, sustainability, and disaster preparedness. This holistic approach strengthens urban resilience, ensuring that cities not only thrive in normal circumstances but also effectively withstand and recover from future hazards.

10.8 Business Models Used in Smart Cities

In this section, the concept of business models commonly utilized in smart cities, is employed to examine their implementation in selected smart city case studies. According to ProptechOS-Smart City Index (Wolniak et al. 2024; Gajdzik and Wolniak 2022), London, Amsterdam, and Berlin are identified as the three European cities best prepared for the future development of smart city initiatives. Through an extensive literature review, this study analyzes the application of these business models within these cities, evaluating their effectiveness and adaptation in real-world urban environments.

10.8.1 Case Studies

10.8.1.1 London

London exemplifies the successful implementation of smart city business models through public-private collaboration, technological innovation, and community engagement. Public-private partnerships (PPPs) play a critical role in infrastructure development, enabling efficient transportation and sustainable energy solutions. Build-operate-transfer (BOT) arrangements further accelerate the deployment of smart city projects by involving private sector investment before transferring ownership to public authorities (Wolniak et al. 2024). Performance-based contracts enhance service efficiency, linking payments to measurable outcomes in areas such as waste management and transportation. Community-centric models, including community land trusts and cooperative housing projects, ensure inclusivity by involving residents in urban development. Innovation hubs like Tech City foster entrepreneurship and technological advancements, driving solutions for urban resilience. Revenue-sharing models optimize service operations by aligning financial incentives with performance metrics, ensuring efficiency and customer satisfaction. Outcome-based financing mechanisms, such as social impact bonds, and channel private capital into addressing social issues like healthcare and homelessness. Additionally, asset monetization strategies, such as leasing public assets, generate revenue for future smart city investments. London's integrated approach to smart city development showcases how innovative financial models, governance strategies, and stakeholder engagement contribute to sustainable and efficient urban growth and resilience while addressing the evolving needs of its residents (Robinson et al. 2021).

10.8.1.2 Amsterdam

Amsterdam's smart city business models integrate public-private collaboration, innovation, and community engagement to promote sustainable urban development and resilience. PPPs drive large-scale projects such as the Zuidas development, combining sustainable infrastructure with public amenities. BOT arrangements, exemplified by the North/South metro line, enable private sector investment in infrastructure, accelerating project completion while reducing financial risks for the public sector (Wolniak et al. 2024). Performance-based contracts enhance service efficiency by linking payments to predefined metrics, particularly in waste management. Community-centric models, including community land trusts and co-housing projects in neighborhoods like Buiksloterham, foster affordable living and social cohesion through resident participation in urban planning and resilience. Innovation hubs, such as the Amsterdam Science Park and Startup Village, support technological advancements by providing collaboration spaces, mentorship, and funding opportunities for startups. Revenue-sharing models optimize public services, including ferries and bicycle rental schemes, aligning financial incentives with efficiency and user satisfaction. Outcome-based financing mechanisms, such as social impact bonds, fund projects addressing homelessness and employment stability. Additionally, asset monetization strategies, including the sale or lease of public land, generate revenue for further smart city initiatives while enhancing resource utilization. Amsterdam's holistic approach ensures inclusive, efficient, and sustainable urban growth and resilience (Mancebo 2020).

10.8.1.3 Berlin

Berlin's smart city business models emphasize public-private collaboration, innovation, and community engagement to drive sustainable urban development and resilience (Wolniak et al. 2024). Public-private partnerships (PPPs) are essential for large-scale projects like the BER Airport, where government and private investors jointly fund, construct, and manage infrastructure, ensuring long-term sustainability and resilience. Similarly, build-operate-transfer (BOT) agreements are widely used in transportation, with private entities financing and operating the S-Bahn and U-Bahn networks before transferring ownership back to the city, facilitating efficient and innovative transit solutions. Performance-based contracts enhance efficiency in sectors like waste management by linking payments to specific performance metrics, ensuring optimal resource utilization. Community-centric models, such as Baugruppen, empower residents to collaboratively design and develop housing projects, fostering social cohesion and inclusivity. Berlin's innovation hubs, including Factory Berlin and the Berlin Technology Park, provide essential resources for startups and entrepreneurs, promoting economic growth and technological advancement. Revenue-sharing models optimize infrastructure operations, particularly in bike-sharing programs, where private operators share profits with the city. Outcome-based financing supports social initiatives like affordable housing, aligning investments with measurable community benefits. Asset monetization strategies, including the sale or lease of public properties, generate funds for further smart city projects while promoting sustainable urban regeneration and resilience (Harrison 2024).

10.8.1.4 Comparison of the Case Studies

This section provides a comparative analysis of smart city business models in London, Amsterdam, and Berlin, highlighting their unique approaches to infrastructure development, sustainability, and community engagement as shown in Table 10.3. Public-private partnerships (PPPs) play a crucial role in all three cities, enabling efficient project delivery and financial risk-sharing. London employs PPPs for transportation and public buildings, Amsterdam integrates them into urban redevelopment projects like the Zuidas district, and Berlin applies them to essential infrastructure, such as airports. Similarly, build-operate-transfer (BOT) arrangements expedite infrastructure deployment, with London focusing on toll roads and airports, while Amsterdam and Berlin utilize BOT models for

Table 10.3 Comparative analysis of smart city business models in London, Amsterdam, and Berlin

Business model	London	Amsterdam	Berlin
Public-private partnerships (PPPs)	Used for transportation and public buildings	Integrated into urban redevelopment (Zuidas)	Applied to essential infrastructure (e.g., airports)
Build-operate-transfer (BOT)	Focus on toll roads and airports	Used for extensive transportation networks	Used for extensive transportation networks
Performance-based contracts	Applied in waste management and resource use	Applied in waste management and resource use	Broad application for service delivery and sustainability
Community-centric models	Affordable housing initiatives	Affordable housing initiatives	Resident-led urban development (Baugruppen)
Innovation hubs	Supports startups and economic growth	Supports startups and economic growth	Supports startups and economic growth
Revenue-sharing models	Applied in transportation systems	Applied in transportation systems	Applied in transportation systems
Outcome-based financing	Linked to affordable housing and urban welfare	Linked to affordable housing and urban welfare	Linked to affordable housing and urban welfare
Asset monetization	Unlocks public asset value for investment	Unlocks public asset value for investment	Unlocks public asset value for investment

Source: Author's compilation

their extensive transportation networks. Performance-based contracts drive efficiency across multiple sectors. London and Amsterdam employ these contracts in waste management and resource utilization, while Berlin applies them broadly to enhance service delivery and sustainability efforts. Community-centric models prioritize inclusivity, with London and Amsterdam fostering affordable housing initiatives and Berlin advancing resident-led urban development projects like Baugruppen. Innovation hubs stimulate economic growth in all three cities, providing essential resources for startups. Revenue-sharing models enhance smart city operations, particularly in transportation systems. Outcome-based financing mechanisms ensure social impact, linking investments to measurable outcomes in affordable housing and urban welfare and resilience. Lastly, asset monetization strategies unlock public asset value, attracting private investment and supporting sustainable urban regeneration and resilience (Wolniak et al. 2024).

10.9 Developing and Implementing Sustainable Urban Resilience Initiatives in Çatalköy–Esentepe, Northern Cyprus

The theoretical framework of Sustainable Urban Resilience Development Initiatives, Green Solutions, Smart Cities, and Industrial Structure can be effectively adapted and implemented in Çatalköy–Esentepe, Northern Cyprus, integrating lessons from global case studies while considering local socio-economic, environmental, and governance contexts.

10.9.1 Contextual Challenges and Opportunities

Çatalköy–Esentepe, a rapidly developing region in Northern Cyprus, faces challenges such as:

- Climate vulnerability (water scarcity, heatwaves, and potential seismic risks).
- Urban sprawl, leading to loss of green spaces and inefficient land use.
- Limited infrastructure resilience, particularly in energy, water, and transportation.
- Economic dependency on tourism and construction, requiring diversification for long-term stability.

However, the region also presents opportunities:

- Strong community engagement and local governance potential.
- Abundant solar energy potential for renewable energy integration.
- Proximity to academic institutions (Eastern Mediterranean University, Near East University) fostering innovation.
- Growing interest in sustainable tourism, aligning with green urban development.

10.9.2 Strategic Implementation Framework

Drawing from **European, Russian, Chinese, and Indian case studies**, a multi-dimensional approach can be adopted:

1. Green Construction and Urban Sustainable Development Planning
    - Adopt green construction standards (energy-efficient buildings, waste reduction, water recycling).
    - Implement Low Emission Zones (LEZs) in high-traffic areas.
    - Nature-based solutions:
        - Urban greening (green roofs, parks, permeable pavements).
        - Stormwater management (bioswales, rain gardens).
        - Restoration of natural habitats (coastal vegetation for erosion control).
2. Smart City and Digital Transformation
    - IoT and Digital Twin Integration:
        - Real-time monitoring of energy use, water consumption, air quality.
        - AI-driven predictive analytics for disaster risk reduction (floods, heatwaves).

- Smart Mobility:
  - E-bike sharing, electric buses, pedestrian-friendly zones.
  - Traffic optimization using AI-based signal systems.
- Resilient Infrastructure:
  - Microgrids and solar-powered street lighting.
  - Decentralized water management (rainwater harvesting, greywater recycling).

3. Governance and Financial Mechanisms
   - Multi-level governance:
     - Local resilience committees (involving municipalities, universities, businesses).
     - Participatory budgeting for community-driven projects.
   - Sustainable financing:
     - Public-Private Partnerships (PPPs) for green infrastructure.
     - Green bonds and EU funding alignment (Horizon Europe, Green Deal initiatives).
4. Economic Diversification and Industrial Resilience (China's Structural Effect)
   - Promote circular economy models:
     - Waste-to-energy plants.
     - Local agro-industry and eco-tourism to reduce construction dependency.
   - Tech and innovation hubs:
     - Collaboration with universities for smart agriculture, renewable energy startups.

10.9.3 Case Study Application: A Pilot Resilience District in Çatalköy-Esentepe

A pilot smart-green neighborhood can be developed, incorporating:

- Energy-efficient buildings (BREEAM/LEED certification).
- Smart grids and solar microgrids.
- Urban farming and green corridors.
- Disaster-resilient infrastructure (earthquake-resistant design, flood barriers).
- Community engagement platforms (digital apps for feedback, emergency alerts).

10.9.4 Monitoring and Evaluation

- KPIs:
  - Reduction in CO_2 emissions (aligned with EU Green Deal).
  - Percentage increase in green cover and renewable energy usage.
  - Economic diversification index (tourism vs. green industries).
- Resilience benchmarking using City Resilience Index (CRI).

By integrating European governance models, green construction, smart city innovations, and disaster resilience strategies, Çatalköy–Esentepe can transition into a sustainable, resilient urban hub. Key success factors include:

- Strong policy alignment (local + international standards).
- Technology-driven solutions (BIM, IoT, AI).
- Community-centric planning (participatory governance).
- Economic resilience (beyond tourism-driven growth).

10.10 Conclusion

This chapter has explored the diverse and evolving landscape of sustainable urban resilience initiatives across European cities and beyond, revealing both the breadth and depth of contemporary approaches to building resilient, inclusive, and environmentally responsible urban environments. Through a rich analysis of case studies and comparative frameworks, it becomes evident that achieving urban resilience is inherently multidimensional—requiring coordinated efforts in governance, infrastructure, finance, technology, and community engagement.

European cities have shown a strong commitment to environmental stewardship and public health, leveraging local governance mechanisms to foster sustainable development. The establishment of low emission zones (LEZs) and the adoption of green infrastructure underscore a strategic pivot toward cleaner air and greener urban ecosystems. These measures are complemented by financial instruments and policy incentives that support sustainable urban growth and ensure long-term viability. The collaborative ethos—manifested through public-private partnerships and citizen co-creation—emerges as a key driver for resilient urban transformation.

International case studies deepen this understanding. Moscow's experience with green construction and urban planning exemplifies the integration of sustainability into post-Soviet cityscapes, supported by emerging green building standards and certification systems in Russia. Similarly, China's ambitious smart city agenda highlights how digital innovation can enhance resilience through improved urban functionality, while India's Smart Cities Mission illustrates the importance of embedding disaster resilience into the DNA of urban development. These global narratives reveal a consistent pattern: resilience is most successful when embedded within broader sustainability strategies and aligned with national priorities.

Furthermore, comparative insights from London, Amsterdam, and Berlin demonstrate how mature European cities are fine-tuning their resilience agendas by leveraging business models that align with local conditions, technologies, and policy goals. These cities illustrate that effective urban resilience is not a one-size-fits-all approach but rather a dynamic process of learning, adaptation, and contextual innovation.

Ultimately, the chapter underscores several key success factors: integrating environmental health into urban planning, ensuring robust governance at local and national levels, enabling financial support mechanisms, promoting technological adoption for smart urbanism, and fostering inclusive, participatory planning processes. As cities face increasingly complex and interconnected risks—from climate change to public health crises—these insights will prove critical for shaping future policies.

In conclusion, sustainable urban resilience is not a destination but an ongoing journey. It demands collaboration, innovation, and a steadfast commitment to shaping cities that are not only smart and efficient but also just livable and prepared for the uncertainties of tomorrow.

References

Abella A, Ortiz-de-Urbina-Criado M, De-Pablos-Heredero C (2017) A model for the analysis of data-driven innovation and value generation in smart cities' ecosystems. Cities 64:47–53

Abusaada H, Elshater A (2023a) Developing a guiding framework based on sustainable development to alleviate poverty, hunger and disease. Archnet-IJAR. https://doi.org/10.1108/ARCH-03-2023-0076

Abusaada H, Elshater A (2023b) Wartime urbanism: three rules for protecting city places. J Urban Int Res Placemaking Urban Sustain

Adreani L, Bellini P, Fanfani M, Nesi P, Pantaleo G (2024) Smart city digital twin framework for real-time multi-data integration and wide public distribution. IEEE Access

Alberti M (2005) The effects of urban patterns on ecosystem function. Int Reg Sci Rev 28:168–192

Allam Z, Dhunny ZA (2019) On big data, artificial intelligence and smart cities. Cities 89:80–91

Bai LM, Xiu CL, Feng XH, Mei DW, Wei Y (2019) Comprehensive assessment of Urban resilience in China and its spatiotemporal differentiation. Study World Geogr 28(77–87):06

Baumol W (1993) Entrepreneurship, management and the structure of payoffs. The MIT Press, Cambridge, pp 1–3

Bourguignon F (2004) The poverty-growth-inequality triangle. The World Bank

Brown L, Greenbaum RT (2017) The role of industrial diversity in economic resilience: an empirical examination across 35 years. Urban Stud 54(6):1347–1366

Caragliu A, Del Bo CF (2019) Smart innovative cities: the impact of smart city policies on urban innovation. Technol Forecast Soc Chang 142:373–383

Chen X, Wei L, Zhang H (2018) Spatial and temporal pattern of urban smart development in China and its driving mechanism. Chin Geogr Sci 28(4):584–599

Clarke NJ, Kuipers MC, Roos J (2020) Cultural resilience and the smart and sustainable city: exploring changing concepts on built heritage and urban redevelopment. Smart Sustain Built Environ 9(2):144–155

Cui L, Yang K (2008) Construction of smart city evaluation index system. Stat Decis Mak 34(33–38):06

Deng T, Zhang K, Shen Z-J (2021) A systematic review of a digital twin city: a new pattern of urban governance toward smart cities. J Manag Sci Eng 6:125–134

Dessouky N, Wheeler S, Salama AM (2023a) Including local actors' perspective in neighborhood sustainability assessment: evidence from Dubai's sustainable city. Smart Sustain Built Environ 13(5):1117–1133

Dessouky N, Wheeler S, Salama AM (2023b) The five controversies of market-driven sustainable neighborhoods: an alternative approach to post-occupancy evaluation. Soc Sci 12(7):367

Doran J, Fingleton B (2018) US Metropolitan Area Resilience: insights from dynamic spatial panel estimation. Environ Plann A Econ Space 50:111–132

Du M, Zhang X, Wang Y, Tao L, Li H (2020) An operationalizing model for measuring urban resilience on land expansion. Habitat Int 102:102206

Egegård CH, Lindborg M, Gren Å, Marcus L, Pont MB, Colding J (2024) Climate proofing cities by navigating nature-based solutions in a multi-scale, social–ecological urban planning context: a case study of flood protection in the city of gothenburg. Sweden Land 13(2):143

EUROCITIES (2014) EUROCITIES strategic framework 2014–2020: towards an EU urban Agenda for cities. https://agendastad.nl/content/uploads/2015/06/EUROCITIES-strategic-framework.pdf

EUROCITIES (2020) Paving the way for sustainable cities. EUROCITIES, vol 13. https://eurocities.eu/latest/paving-the-way-for-sustainable-cities/. Accessed 10 Aug 2023

EUROCITIES (2025) Towards zero-emission mobility in Europe. https://eurocities.eu/latest/towards-zero-emission-mobility-in-europe/

European Commission (2014) European green capital city. European Commission. https://environment.ec.europa.eu/topics/urban-environment/european-green-capital-award/about-awards_en#background

European Environment Agency (2024) Urban sustainability. https://www.eea.europa.eu/en/topics/in-depth/urban-sustainability

Evans R, Karecha J (2014) Staying on top: why is Munich so resilient and successful? Eur Plan Stud 22:1259–1279

Federal Law (2009a) Federalniy zakon Rossiyskoi Federatsii 'Tekhnicheskii reglament bezopasnosti zdanii i sooruzhenii' ot 30.12.2009 N384-FZ [The Federal Law of the Russian Federation 'Technical regulations of the buildings and structures safety' from December 30, 2009, No.384 FZ]. https://www.consultant.ru/document/cons_doc_LAW_95720/. Accessed 25 Aug 2016

Federal Law (2009b) Federalniy zakon Rossiyskoi Federatsii 'O vnesenii izmeneniy v Federalniy zakon 'O tekhnicheskom regulirovanii' ot 30.12.2009 N385-FZ' [The Federal Law of the Russian Federation 'On inserting changes into the Federal Law "On technical regulations" No.385-FZ' from December 30, 2009]. https://www.consultant.ru/document/cons_doc_LAW_95721/. Accessed 25 Aug 2016

Fingleton B, Palombi S (2013) Spatial panel data estimation, counterfactual predictions, and local economic resilience among British towns in the Victorian Era. Reg Sci Urban Econ 43(4):649–660

Gajdzik B, Wolniak R (2022) Framework for R&D&I Activities in the Steel Industry in Popularizing the Idea of Industry 4.0. J Open Innov Technol Mark Complex 8:133

García-Blanco G, Navarro D, Feliu E (2023) Adopting resilience thinking through nature-based solutions within urban planning: a case study in the city of València. Buildings 13(5):1317

Gkontzis AF, Kotsiantis S, Feretzakis G, Verykios VS (2024) Enhancing urban resilience: smart city data analyses, forecasts, and digital twin techniques at the neighborhood level. Future Internet 16(2):47

Grossman GM, Krueger B (1991) Environmental impacts of a north American free trade agreement. NBER Working Paper

Harrison AL (2024) Feral surfaces: building envelopes as intelligent multi-species habitats. Archit Des 94: 38–45

Henderson H, Sanquiche R, Nash T (2015) Breakdowns driving breakthroughs: 2015 Green Transition Scoreboard® report. Ethical Markets Media, Saint Augustine, FL

Hu Q, Zheng YP (2020) Smart city initiatives: a comparative study of American and Chinese cities. J Urban Aff

Huang D, Chan RCK (2018) On 'land finance' in urban China: theory and practice. Habitat Int 75:96–104

IBC (2009) International Building Code® 2009. International Code Council, Inc. https://www.iccsafe.org/products-and-services/i-codes/ibc/

ICLEI (2025) Resilience and Climate Adaptation, International Council for Local Environmental Initiatives, Local Governments for Sustainability, Europe. https://iclei-europe.org/topics/resilience-and-climate-adaptation/

IRADe (2019) Sustainable and Disaster Resilient Urban Development Assessment of 10 Cities in India, https://irade.org/Sustainable%20and%20Disaster%20Resilient%20Urban%20DevelopmentFinal%20.pdf. Accessed 17 Mar 2025

Ismagilova E, Hughes L, Dwivedi YK, Raman KR (2019) Smart cities: advances in research—an information systems perspective. Int J Inf Manag 47:88–100

Jang FT, Li XP (2018) Evolution and development of China's industrial policy in the 40 years of reform and opening up: and on the transformation of China's industrial policy system. Manage World 34(10): 73–85

Kitchin R (2014) The real-time city? Big data and smart urbanism. GeoJournal 79:1–14

Kramer J, Silverton S, Späth P (2024) Urban governance arrangements for sustainability and justice–linking theory with experience. Urban Transform 6(1):6

Kummitha RKR, Crutzen N (2017) How do we understand smart cities? An evolutionary perspective. Cities 67:43–52

Labaka L, Maraña P, Giménez R, Hernantes J (2019) Defining the roadmap towards city resilience. Technol Forecast Soc Chang 146:281–296

Lee S, Kim DH, Nam I (2015) The impact of mobile broadband infrastructure on technological innovation. Int Telecomm Policy Rev 22(2):93–108

Leitner H, Sheppard E, Webber S, Colven E (2018) Globalizing urban resilience. Urban Geogr 39(8):1276–1284

Mai X, Zhan C, Chan RCK (2021) The nexus between (re)production of space and economic resilience: an analysis of Chinese cities. Habitat Int 109:102326

Majewska A, Denis M, Jarecka-Bidzinska E, Jaroszewicz J, Krupowicz W (2022) Pandemic resilient cities: possibilities of repairing Polish towns and cities during COVID-19 pandemic. Land Use Policy 113:2022

Manca AR, Benczur P, Giovannini E (2017) Building a scientific narrative towards a more resilient EU society. JRC Science for Policy Report, Publications Office of the European Union, Brussels. https://publications.jrc.ec.europa.eu/repository/handle/JRC106265. Accessed 11 Mar 2025

Mancebo F (2020) Smart city strategies: time to involve people. Comparing Amsterdam, Barcelona and Paris. J Urban 13:133–152

Margarita A (2014) Smart city policies: a spatial approach, cities 41

Martin R, Sunley P, Gardiner B, Tyler P (2016) How regions react to recessions: resilience and the role of economic structure. Reg Stud 50:561–585

Matthies F, Menne B (2009) Prevention and management of health hazards related to heatwaves. Int J Circumpolar Health 68(1):8–22

Meduri Y, Singh R, Manoharan G (2024) Key networks to create disaster resilient Smart Cities Mission: a case for remodeling India's Smart Cities Mission to include disaster resilience. Resil Cities Struct 3(4):52–62

Michaels G, Rauch F, Redding SJ (2012) Urbanization and structural transformation. Q J Econ 127(2):535–586

National Standard (2013) Natsionalnii standart Rossiyskoi Federatsii 2013. GOST R 54964-2012. 'Otsenka sootvetstviya. Ekologicheskiye trebovaniya k ob'ektam nedvizhimosti'. Moskva, Standartinform [National Standard of the Russian Federation. GOST K 54964-2012. 'Compliance Assessment. Environmental Requirements to Real Estate'. Federal Agency for Technical Regulation and Metrology. Moscow, Standardinform]

Osei-Kyei R, Tam V, Komac U, Ampratwum G (2023) Critical review of urban community resilience indicators. Smart Sustain Built Environ

Peres E, du Plessis C, Landman K (2017) Unpacking a sustainable and resilient future for Tshwane. Procedia Eng 198:690–698

Plummer R, Baird J, Bullock R, Dzyundzyak A, Dupont D, Gerger Swartling A, Johannessen A, Huitema D, Lyth A, de Lourdes Melo Zurita M, Munaretto S, Smith T, Thomsen D (2018) Flood Governance: a multiple country comparison of stakeholder perceptions and aspirations. Environ Policy Gov 28(2):67–81. https://doi.org/10.1002/eet.1796

Porfiriev B, Dmitriev AN, Vladimirova IL, Tsygankova AA, Guryev VV (2016) 'Zeleniye' standarti:Otsenka sostoyaniya i zadachi sovershenstvovaniya normativnoi bazi [Green standards:Assessment of the existing state and the tasks for improvement of the regulatory framework]. Standarti i Kachestvo [Standards and Quality] 8:16–21

Porfiriev BN, Dmitriev A, Vladimirova I, Tsygankova A (2017) Sustainable development planning and green construction for building resilient cities: russian experiences within the international context. Environ Hazards 16(2):165–179. https://doi.org/10.1080/17477891.2017.1280000

RCN (2022) What is urban resilience? Resilient Cities Network. https://resilientcitiesnetwork.org/what-is-urban-resilience/. Accessed 10 Aug 2023

Resilient Cities Network (2025) NetZero cities—supporting European cities in their transition to net-zero emissions by 2030. Resilient Cities Network, https://resilientcitiesnetwork.org/netzerocities/

Ribeiro PGJ, Pena Jardim Gonçalves LA (2019) Urban resilience: a conceptual framework. Sustain Cities Soc 50:101625

Robinson J, Harrison P, Shen J, Wu F (2021) Financing urban development, three business models: Johannesburg, Shanghai and London. Prog Plan 154:100513

Salama AM, Patil MP, MacLean L (2024) Urban resilience and sustainability through and beyond crisis–evidence-based analysis and lessons learned from selected European cities. Smart Sustain Built Environ 13(2):444–470

Shi D, Ding QH, Wei P, Liu J (2018) Can smart city construction reduce environmental pollution. China Ind Econ 6:117–135

Slaev AD, Nedovic-Budic Z (2017) The challenges of implementing sustainable development: the case of Sofia's master plan. Sustainability 9(1):15

Soyata T, Habibzadeh H, Ekenna C, Nussbaum B, Lozano J (2019) Smart city in crisis: technology and policy concerns. Sustain Cities Soc 50:101566

Tommi I (2018) Reflections on the innovative city: examining three innovative locations in a knowledge bases framework. J Open Innov Technol Mark Complex 1(1)

Tong Y, Ma Y, Liu HM (2020) Short-term impact of COVID-19 on urban migration in China and evaluation of urban resilience. Acta Geograph Sin 75(11):2505–2520

UNDRR (2024) The ten essentials for making cities resilient. https://mcr2030.undrr.org/ten-essentials-making-cities-resilient. Accessed 17 Mar 2025

UN-Habitat (2016) World cities report 2016: urbanization and development: emerging futures. United Nations Human Settlements Program—UN Habitat, Nairobi

Urban Planning Code of the Russian Federation (2004). Urban Planning Code of the Russian Federation dated 29.12.2004 N 190-FZ. https://www.iibh.org/kijun/pdf/Russia_01UrbanDevelopmentCode_E.pdf

Wang M, Li YF, Ma SC (2020) Whether the construction of smart cities promotes the upgrading of industrial structure. Financ Sci 12:56–71

Wang SJ, Yi SJ (2018) An empirical study on urban competitiveness of northeast China based on population agglomeration. Population Journal 40:18–29, 03

Weil C, Bibri SE, Longchamp R, Golay F, Alahi A (2023) Urban digital twin challenges: a systematic review and perspectives for sustainable smart cities. Sustain Cities Soc 99:104862

WHO (2022a) Urban planning for resilience and health: key messages. World Health Organization. https://www.who.int/europe/publications/i/item/WHO-10665-355760. Accessed 12 Feb 2025

WHO (2022b) Urban planning for health—experiences of building resilience in 12 cities, Second report on protecting environments and health by building urban resilience, World Health Organization, Regional Office for Europe. https://iris.who.int/handle/10665/355762. Accessed 17 Feb 2025

WHO EHCN (2018) World Health Organization—copenhagen consensus of mayors: healthier and happier cities for all: a transformative approach for safe, inclusive, sustainable and resilient societies, World Health Organization. Regional Office for Europe, Copenhagen

WHO EHCN (2019) WHO European Healthy Cities Network Phase VII (2019–2024): support package for implementation compendium of tools, resources and networks, World Health Organization, European Healthy Cities Network, World Health Organization. Regional Office for Europe, Copenhagen

Wolniak R, Gajdzik B, Grebski M, Danel R, Grebski WW (2024) Business models used in smart cities—theoretical approach with examples of smart cities. Smart Cities 7(4):1626–1669

Xia X, Zhou Z, Hou YL, Zhang YL, Q X, Zhang XJ (2020) Evolutionary game analysis and simulation of the influencing factors of benefit distribution in the construction of new smart city. Journal of Technology Economics 39:59–65+85, 04

Zhou Q, Zhu M, Qiao Y, Zhang X, Chen J (2021) Achieving resilience through smart cities? Evidence from China. Habitat Int 111:102348

Zhu S, Li D, Feng H (2019) Is smart city resilient? Evidence from China. Sustain Cities Soc 50:101636

11 Conclusion

The quest for smart city development is gaining momentum globally. This is especially true for countries that are scientifically and environmentally sophisticated. This chapter explores overviews of smart cities, origins, importance, elements, issues, and future directions. The chapter sets the scene by putting urbanization from the Second Industrial Revolution to the present from a context regarding the development of the human population and innovation. Therefore, the setting for the emergence of smart cities is linked to this historical foundation and the fact that urban issues have increasingly become more complex, with climate change, infrastructure needs, and economic disparities as the embodiment.

Chapter 2 discusses different tasks, including IBM data-driven operational optimization, sustainability, and inclusivity. This definitional inconsistency is acknowledged as a strength and a weakness since it may lead to localized adaptation. However, from a global perspective, the chapter shows various ways of taking innovative city initiatives. Examples of the implementation of Singapore's Smart Nation framework on one end and Amsterdam's sustainability-minded efforts on the other from case studies in Asia, Europe, North America, Africa, and Oceania are presented. While these case studies offer interesting lessons, they also reveal that not all countries have the same level of access to technology, governance models, and financial capabilities, a disparity that, while useful as we work to broaden smart city projects to lower-developed regions, also complicates scalability and replicability.

On the key components of smart cities, Chap. 2 shows infrastructure, governance, technology integration, and sustainability. Moreover, it demonstrates that technological advancements will never allow us to create truly smart cities without supportive policy frameworks and regulations. However, the governance discussion is somewhat hypothetical since it assumes that participatory governance models do not have political and bureaucratic constraints associated with them.

Also, highlighting the challenges of smart cities, the study shows that these challenges include data privacy, cybersecurity threats, the digital divide, financial constraints, and other matters of scalability and interoperability. In this, they investigate this in-depth, whilst a more critical debate into how corporate interests and privatization affect the development of smart cities could further imply that such development will act to perpetuate, rather than solve, socio-economic inequalities.

Additionally, Chap. 2 shows the future trends and directions, focusing on next-generation technologies, including 5G, blockchain and AI, and the smart city system's view in facilitating innovation. The chapter convincingly claims that cities must adapt to global crises like floods and pandemics but does not fully examine the moral and legal problems associated with the rise of surveillance and automation within urban spaces. Chapter 2 concludes with a narrative on the evo-

V. Yorucu et al., *Modern Approach of Resilient and Sustainable Smart Cities*, Sustainable Development Goals Series, https://doi.org/10.1007/978-3-032-05715-0_11

lution and implementation of smart cities. However, there is an element of optimism on whether or not they have transformative potential. While it acknowledges the challenges, it underplays the complexity of political and economic barriers to equitable smart city development. Enhancing the analysis and a more nuanced understanding of the future of urbanization would be to engage more critically with power relations and corporate governance in influencing smart city agendas.

Chapter 3 focuses on data-driven governance, which is crucial to improving decision-making and public services in cities. This highlights the significance of smart city governance across countries with the emergence of smart cities. While smart governance is a priority in developing smart cities and nations, its obstacles and tactics remain unclear. The role of governance in the development and evolution of smart cities is examined critically. Chapter 3 also argues that governance is not an enabler effect of a smart city but a precondition of a smart city project. Chapter 3 begins with an introduction to smart cities by addressing definition shortcomings and two opposing views of technological solutions and people's approaches. Despite this, the current governance of technologies does not have a universally accepted framework, which makes balancing the need for technologies with development and social needs infeasible.

The study then examines the structural dimensions of smart city governance, identifying several elements such as the policy frameworks, the decision-making processes, the regulatory mechanisms and the collaboration strategies. It takes a critical look at increasing data deciding with the aid of data, while real-time data is a higher performance. Also, it is a risk to privacy, security and democratic participation. The discussion also calls out the practice of many smart city schemes assuming technological infrastructure ahead of social equity, thereby deepening the existing digital divide or making the disadvantaged community even more marginalized.

Governing models used in smart cities, from centralized to decentralized and hybrid models, are also a major theme of Chap. 3. To criticize the centralized governance mechanism with a top-down structure that, while fast in making decisions, can be inflexible, exclusive of its own, and lacks security from hackers. On the other hand, decentralized models are commended for driving community involvement and localized solutions while being lambasted for possible inefficiencies, lack of coordination, and scalability issues. Hybrid governance is presented as a new solution that includes both characteristics, but with monopoly being controlled and flexibility applied.

It deliberates stakeholders' roles (government agencies, private sector entities, civil society organizations, among others) critically. Furthermore, Chap. 3 warns against corporate overreach, where private interests take over the governance of cities whose peoples' names are invoked to cheer smart city success as a manifestation of necessary multi-stakeholder collaboration. Similarly, public participation is a declared goal yet rarely realized, caused by barriers such as digital illiteracy, lack of transparency, and tokenistic participation process.

The chapter then examines public issues such as data privacy rights, unavoidable interoperability, and financial constraints. Talking smart cities, it claims that such smart cities could be useful for economic and environmental growth, yet implementation of smart cities seldom emphasizes inclusiveness; instead, it emphasizes efficiency. Additionally, regulatory frameworks are disconnected, hampering the development of consistent governance practices in various regions.

The case studies of successful smart city governance models, such as those in Singapore, Barcelona, and Copenhagen, shed light on how this technology can be harnessed to achieve a balance of technological innovation and the well-being of the citizens. Nevertheless, the chapter is critical of the sustainability and scalability of these models as not all models developed can be replicated effectively in different socio-economic contexts. Also, the chapter advocates for adaptive, participatory, and ethical smart city governance, focusing on technology, robust regulatory oversight, social inclusivity, and long-term sustainability rather than a techno-utopian approach.

Chapter 4 explores smart city development dynamics in developed and developing economies. Firstly, it argues that within the context of rapid global urbanization, smart city models are needed first, and these initiatives all promote sustainability, governance, and quality of life, but the success of which depends on the availability of economic status, infrastructure and policy. Cities such as Singapore, London, and Amsterdam already use AI, IoT, and big data, but critics say these models exacerbate digital divides and raise privacy and data control issues.

Also, smart city initiatives are difficult to implement in developing economies in Africa, South Asia and Latin America. The chapter rejects reactive and fractured approaches to smart city projects in territories with political agendas and external funding. Infrastructure deficiencies, reliance on foreign investment, and a digital divide hinder scalability. Some cities like Nairobi and Kigali use mobile-first digital solutions, but they have not addressed the systemic governance and lack of infrastructure.

This comparison also points to a disparity between the city implementation of smart cities in developed and developing economies. Under developing economies, one devised innovative solution at frugal costs to solve the urban problems, while the developed economies developed data-driven governance, sustainability and automation. On the other hand, the latter is politically and financially unstable; with unstable regulation, corruption is a good native value that is impossible. The chapter critiques the assumption that smart city models from developed countries that follow unthinkingly can be transplanted onto developing economies without any adjustments, and local conditions, social equity, and government capacity need to be considered.

It then goes into emerging technologies that will determine the future of smart cities, such as AI, Blockchain, 5G, and IoT. The chapter criticizes the ethical concerns of these innovations for giving up privacy rights in surveillance-rich environments in such models. It cautions against tech optimism running wild and advocates models of governance based on ethical AI, cybersecurity, and including citizens. Chapter 4 concludes by showing that smart cities require adaptive governance, multi-stakeholder participation and equitable access to smart cities' benefits. Rather, it presses the developing economies to neglect the high-tech strategies of rich countries and embrace scalable, inclusive models. Ultimately, it demonstrates that smart cities will be successful not only if they are driven by technology but also when robust policies, participatory governance, and development frameworks related to sustainable development promote the welfare of humans rather than just digital transformation.

Chapter 5 deals with the discussion on rapid population growth and its negative impact on the urban city. As a case study, Catalkoy and Esentepe villages have been chosen to examine the quality of life, welfare level, safety, security and resilience of the region. Besides that, there is serious environmental problem, which is caused by Teknecik electricity plant. It is endangering human life due to the usage of fossil fuel (mostly fuel oil and diesel) and therefore needs to be substituted with other energy sources rather than fossil fuel turbines.

To make Catalkoy-Esentepe a smart city, smart energy meters should be installed all over the place and letting it be used. Demand for energy can then be easily assessed; costs can be lowered and pollution caused by energy consumption can be reduced. The island geographically sits in a very strategic location, enjoying plenty of sunny days, and therefore photovoltaic parks should be expanded all over the region to be relieved from Teknecik Power Plant. Being connected to a grid system, energy storage is a serious constraint that one should not ignore. The other alternative way to resolve the energy supply problem without Teknecik is to be connected to sub-sea electricity interconnector gridlines with Turkey. If ever such a project succeeded, the cost of energy will be reduced in exchange for using much cleaner energy.

The other way to reduce carbon emissions in urban cities is to encourage public transport. The city must have alternative ring roads, bypasses, underpasses, and elevated roads. The repair and maintenance of village roads are needed to advance road trustworthiness.

For the sake of good water management, a smart city must install smart water meters to assess the water consumption by the users. Water leakages in the distribution system can be prevented by installing Scada systems. An early warning system through Scada network can prevent such water losses. The Ayandare and Uzundere water dams in Catalkoy-Esentepe have great potential both for the city's economic expansion, especially for tourism and agriculture industries. Both dams are under care and regularly rehabilitated to let the water reservoir be well utilized during hot summer months and drought.

The solid waste management system in Catalkoy-Esentepe, and the collection of waste of electric and electronic equipment (WEEE), reuse or exporting it to the Southern Cyprus for recycle is also important for the future of smart city of Catalkoy-Esentepe. The Municipal Solid Waste Strategy plan needs to be prepared and implemented urgently.

In smart cities, ICT facilities must be available and commonly used by all residents. A well-developed high technology internet with 5G infrastructure equipped with cell phones towers should be available all over in Catalkoy-Esentepe, to make the city run as a smart city. The digital transformation of the city and providing digital municipal services as well as good governance structure through online network is essential.

Spatial planning is essential for the development of better social life in Catalkoy-Esentepe, which is summarized in Chap. 5.

Chapter 6 concentrated on how a traditional city can be transformed into a safe, inclusive, resilient and sustainable smart urban city. As a micro-level study, Catalkoy-Esentepe region in the Northern part of Cyprus have been chosen as case study to investigate the historical, cultural and economic impacts of the region before implementing necessary action plans, policies, strategies and objectives to achieve the transformation process. Both Catalkoy and Esentepe have rich national and historical sites to promote the region as a sustainable tourist region. After careful investigation, we noticed that the region holds almost 50% of five-star tourist accommodation and lodging facilities and enrolls a significant amount of labor in the service industry. Additionally, there is a large investment incentive program for the development of capacity building in the tourism sector through new hotels and other types of tourism activities. The incentive program of North Cyprus, which is administered by the State Planning Organization and a certificate of approval provides exemption from income tax for a stipulated number of years, and duty-free imports and other tax privileges to foreign and domestic investors. Other than tourism, the construction sector boomed in the region during the last few decades and the estate market became an important source of foreign exchange income. Besides that, the region has prosperous employment opportunities. Such economic developments have triggered urban mobility, which initiated rapid population growth, increasing demand for housing and labor. Urban sprawl problems have negative externalities to city life, such as traffic congestion, pollution, increasing rents, insufficient housing, lack of utility services, and infrastructure. The mortality rate has increased in the region over the last two decades because of Teknecik Power Plant. The historical, cultural and economic impact of city life has attracted rapid urban mobility due to increasing potential of the city, in terms of economic benefits and employment. Chapter 6 shed light on the basic urgent needs of the city and necessitated action plans to make the city safer and more resilient.

In Chap. 7, the implementation of spatial planning, action plans, objectives and policy recommendations and the assessment needed to succeed the transformation process has been analyzed, which focused on the UN's sustainability agenda and smart city initiatives. Catalkoy-Esentepe villages have challenges in tourism and construction industries. Development in the real estate sector has already caused rapid urbanization. There is a large amount of investment to the property market, which underpinned by rising income levels and marked improvements in urban infrastructure and utility services. We wonder whether construction-led growth leads to economic growth in the region or foreign capital flow leads to construction growth. Perhaps, the economic growth through tourism leads to construction growth, which caused significant amounts of foreign capital flow in Catalkoy-Esentepe region. Nonetheless, urban development proceeded in the absence of environmental planning and a coherent approach of housing. Rapid urbanization development in Catalkoy-Esentepe has resulted in unplanned urban sprawl, substandard and unaffordable housing, traffic congestion, air and sound pollution, infrastructure bottlenecks, blackouts, unfriendly urban design, green spaces and waste management problems. Moreover, Chap. 7 focuses on key performance indicators that are required to assess the city's successful performance.

The development of smart and resilient cities represents a transformative approach to urban planning and governance, addressing the growing challenges of modern urbanization, which was examined in Chap. 8. These cities integrate cutting-edge technologies, such as AI, IoT, and big data, with strategies for disaster resilience to create urban environments that are not only efficient and adaptive under normal circumstances but also robust in the face of crises. The concept of a smart resilient city bridges the gap between traditional smart cities and disaster-resilient urban frameworks, fostering seamless operations and rapid recovery.

At the core of this approach is the combination of technological innovation and resilience-building measures. Smart cities contribute data-driven tools, real-time analytics, and connected infrastructures, while resilient cities enhance governance structures, emergency preparedness, and recovery capabilities. Together, they create a dynamic urban model capable of proactive risk assessment, effective disaster response, and long-term adaptability. This integrated model goes beyond post-disaster recovery by emphasizing early warning systems, predictive decision-making, and continuous learning from past events.

The significance of smart resilient cities lies in their ability to address critical needs in disaster prevention, urban governance, and resource allocation. They not only safeguard public safety during emergencies but also improve the quality of life for citizens by ensuring efficient public services, sustainable resource management, and equitable access to infrastructure. Furthermore, the integration of AI into scenario planning enhances the adaptability of urban systems, enabling cities to predict and mitigate the impacts of natural and human-made disasters more effectively.

In conclusion, smart and resilient cities offer a comprehensive, technology-driven framework for addressing the complexities of urbanization. By combining intelligence and resilience, they pave the way for a sustainable, inclusive, and adaptive future, transforming urban areas into hubs of innovation and security for generations to come.

The framework for territorial spatial resilience planning and construction provides a comprehensive approach to strengthening the adaptability and robustness of urban and regional landscapes. It emphasizes three key dimensions: technical methodologies, spatial governance, and operational systems, ensuring that resilience becomes an integral part of land and urban planning processes.

The primary objectives of resilience planning include the development of national and regional systems that can effectively prevent, respond to, and recover from disasters. A significant transition is proposed, shifting from traditional engineering-based disaster prevention to governance-based resilience frameworks. The inclusion of disaster scenario planning in spatial resilience strategies provides a multi-scale perspective, offering robust theoretical and practical foundations for urban and territorial emergency management.

The "one blueprint" approach ensures holistic planning by harmonizing multiple dimensions of spatial planning—disaster prevention, environmental management, and development strategies—into a unified system. At the technical level, emerging technologies such as big data analytics and real-time monitoring are integrated to improve decision-making, public participation, and disaster management. Spatial governance is enhanced through a balance of binding regulations and adaptive guiding requirements, optimizing resource allocation and ensuring sustainable development.

This roadmap also emphasizes collaborative multi-objective planning, ensuring coherence across resilience, ecological, and economic goals. Aligning resilience strategies with broader initiatives like carbon neutrality and green urban development fosters long-term sustainability. Ultimately, this framework provides a structured and dynamic approach to territorial resilience, safeguarding public safety, enhancing urban functionality, and supporting sustainable development in the face of increasing risks and uncertainties.

Chapter 9 examined urban resilience, which is the capacity of urban systems, including socio-ecological and socio-technical networks, to maintain or quickly restore functions and adapt to changes. It encompasses four pillars: recovery, adaptation, transformation, and resilience. Urban resilience can be viewed from engineering, ecological, and socio-ecological perspectives. Smart cities and resilient cities share development trajectories, though not all smart cities are resilient to unexpected events. Urban areas face challenges from rapid urbanization, inadequate planning, and natural hazards, particularly in emerging economies. Resilience and sustainability are critical strategies for improving urban planning and addressing hazards, highlighted in the Sustainable Development Goals (SDG 11). Understanding resilience involves exploring its theoretical evolution and interdisciplinary applications.

Integrating Urban Digital Twins (UDT) into smart city frameworks has the potential to revolutionize urban development. The applications of UDT in urban planning are diverse, offering innovative solutions for energy management, infrastructure monitoring, and public services, among other areas. Many cities are exploring the concept of UDTs to enhance urban administration, development, and decision-making processes. While a fully realized city-scale UDT has not yet been achieved, there is a clear movement toward this goal. Even though urban digital twins are still in the early stages of development, they can already contribute to improvements in urban management. A city digital twin allows for the integration of city planning and management into a single platform, eliminating the need for multiple management systems. UDT can enable faster response times through autonomous actions and enhance city management efficiency by providing real-time insights into the city's condition.

Sustaining resilience in the AEC industry aligns with environmental responsibility and Construction 5.0's vision of smart, sustainable structures. This includes adopting sustainable techniques, decentralizing production, and large-scale customization. Success relies on balancing economic, environmental, and social aspects, with blockchain aiding sustainability and traceability. Society 5.0 emphasizes energy efficiency, automation, and occupant comfort in smart built environments.

Chapter 10 explored the diverse and evolving landscape of sustainable urban resilience initiatives across European cities and beyond, revealing both the breadth and depth of contemporary approaches to building resilient, inclusive, and environmentally responsible urban environments. Through a rich analysis of case studies and comparative frameworks, it becomes evident that achieving urban resilience is inherently multidimensional—requiring coordinated efforts in governance, infrastructure, finance, technology, and community engagement.

European cities have shown a strong commitment to environmental stewardship and public health, leveraging local governance mechanisms to foster sustainable development. The establishment of low emission zones (LEZs) and the adoption of green infrastructure underscore a strategic pivot toward cleaner air and greener urban ecosystems. These measures are complemented by financial instruments and policy incentives that support sustainable urban growth and ensure long-term viability.

International case studies deepen this understanding. Moscow's experience with green construction and urban planning exemplifies the integration of sustainability into post-Soviet cityscapes, supported by emerging green building standards and certification systems in Russia.

Similarly, China's ambitious smart city agenda highlights how digital innovation can enhance resilience through improved urban functionality, while India's Smart Cities Mission illustrates the importance of embedding disaster resilience into the DNA of urban development. These global narratives reveal a consistent pattern: resilience is most successful when embedded within broader sustainability strategies and aligned with national priorities.

Furthermore, comparative insights from London, Amsterdam, and Berlin demonstrate how mature European cities are fine-tuning their resilience agendas by leveraging business models that align with local conditions, technologies, and policy goals. These cities illustrate that effective urban resilience is not a one-size-fits-all approach but rather a dynamic process of learning, adaptation, and contextual innovation.

Ultimately, Chap. 10 underscores several key success factors: integrating environmental health into urban planning, ensuring robust governance at local and national levels, enabling financial support mechanisms, promoting technological adoption for smart urbanism, and fostering inclusive, participatory planning processes. As cities face increasingly complex and interconnected risks—from climate change to public health crises—these insights will prove critical for shaping future policies.

In conclusion, sustainable urban resilience is not a destination but an ongoing journey. It demands collaboration, innovation, and a steadfast commitment to shaping cities that are not only smart and efficient but also just livable and prepared for the uncertainties of tomorrow.